The Glue Book

The
Glue Book

WILLIAM TANDY YOUNG

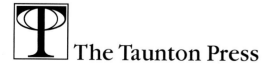

The Taunton Press

Cover photos by William Tandy Young

BOOKS & VIDEOS

for fellow enthusiasts

Text ©1998 by William Tandy Young
Photographs ©1998 by The Taunton Press, Inc., except where noted
Illustrations ©1998 by The Taunton Press, Inc.

Printed in the United States of America
10 9 8 7 6 5 4 3 2 1

The Taunton Press, Inc., 63 South Main Street,
PO Box 5506, Newtown, CT 06470-5506
e-mail: tp@taunton.com

Distributed by Publishers Group West

Library of Congress Cataloging-in-Publication Data

Young, William Tandy.
      The glue book / William Tandy Young.
          p.   cm.
      Includes index.
      ISBN 1-56158-222-0
      1. Glue. 2. Adhesives. I. Title.
TP968.Y68   1999
668'.3—dc21   98-26894
                                 CIP

About Your Safety

Working wood is inherently dangerous. Using hand or power tools improperly or ignoring standard safety practices can lead to permanent injury or even death. Don't try to perform operations you learn about here (or elsewhere) unless you're certain they are safe for you. If something about an operation doesn't feel right, don't do it. Look for another way. We want you to enjoy the craft, so please keep safety foremost in your mind whenever you're in the shop.

This book is dedicated to my family.

Acknowledgments

Anyone who says writing is a lonely profession has never published a book about glue. During the course of writing this book, I was continually sustained by the help and encouragement of scores of people who took an active interest in what I was doing and who firmly believed in the value of it. My thanks go out to them as follows:

Helen Albert, Jonathan Binzen, Allan Breed, Cathleen Calmer, Melissa Carr, Brian Considine, Kim Cristiano, Kristie DePrete, Ruth Dobsevage, Michael Dresdner, John Driggers, Eric Englander, Susan Fazekas, Patricia C. Fitzmaurice, David Frechtman, Paula Garbarino, Zack Gaulkin, Letitia Hafner, Boyd Hagen, Judith Hanson, B. A. Harrington, Georgette Iannuzzi, Jeff Jewitt, Thomas Johnson, Silas Kopf, Peter Korn, Maggie Lange, Vincent Laurence, Andrew Millican, Richard Scott Newman, Terrie Noll, David Orth, Jere Osgood, Ken Parker, Lance Patterson, Russell Perry, Scott Phillips, Jennifer Renjilian, Joanne Renna, Kim Schmahmann, Scot Schmidt, Douglas Schroeder, Harry Sleeper, Tray Sleeper, Carl Swensson, Christine Thomson, Jim Tolpin, Paul Tuller, Tania Wilcke, and Amelia Young.

The following people provided literature, samples, and all-around expertise:

Andy Allen (Lie-Nielsen Toolworks); Dick Anderson and Mike Richardson (System Three); Jon Behrle, Tim Rinehart, and Greg Plum (Woodcraft); Richard Cook, Ed McCue, and Ken Blake (National Casein); Karen Cox (Nelson Paint Co.); J. B. Currell and Tony De Lima (MAS Epoxies); Elaine Fairchild (Rogers & Co.); Dave Falkanhagan and Roberta DeDamos (BALSA USA); Chris Filardi (Bordens, Inc.); Jim Godfrey (Forest Products Research); Professor R. Bruce Hoadley (University of Massachusetts); Darryl Keil (Vacuum Pressing Systems); Brian Knight (Gougeon Brothers); Russ Kopp (H.B. Fuller); Bob Jacques (Spotlite); Richard Jones (CMI Ingredients); Becky Martin and Donna Russell (3M); Teri Masaschi (Woodworker's Supply of New Mexico); Sammy Mayeux (AmBel/Excel); Bob, Eric, Tim, and Richard Norland (Norland Products); Kyle Ondricek (University Products); Kent, Jeff, and Brian Pitcher (Custom Pak Adhesives); Paul Ross (Ross Adhesives); Dr. Terry Sellers (Mississippi State University), Steve Smith and Emil Munkres (Smith & Co.); John Starer (Adhesive Technologies); Eugene Thordahl (Bjorn Industries); Martin Torbert (National Starch); Jacob Utzig (Milligan and Higgins); Mark Williams (Veritas Tools); and Dale Zimmerman, Mike Syfert, and Lynne Jeffrey (Franklin International).

Special thanks are due to Chris Minick, chemist, author, and woodworker, a fellow whose knowledge is matched only by his perspective, sense of humor, and helpfulness. I've lost track of the number of things he has revealed to me and the number of times he has rescued me from my misconceptions.

My deep appreciation goes to my friend and shop partner Joseph Twichell, who endured this project without ever complaining. Joe kept the shop going in my absence, took photos and also appeared in them, worked in the dark when the shop lights were off during photo shoots, and continually made room for lights and cameras regardless of what he was in the midst of doing.

Finally, my deepest and most heartfelt gratitude goes to my wife, Mary, and my daughters, Amelia, Charlotte, and Eliza, for their infinite patience and for the untold sacrifices that they made on my behalf as I wrote this book. If you want to know the true weight of an entablature, just ask the pillars.

I can't properly express my indebtedness to all these people by simply naming them. Owing so much to so many is a profound and enriching responsibility.

Contents

Introduction   4

1   Understanding Glues and Adhesives   6
Chemical Basis and Behavior   6
Formulation   10
Handling and Storage   12
Performance   14

2   Natural Glues   20
Chemical Basis and Behavior   20
Plant-Based Glues   22
Fish Glue   23
Hide Glue   25
Rabbit-Skin Glue   31
Bone Glue   31
Casein Glue   31

3   PVA Glues   34
Chemical Basis and Behavior   36
Formulation   36
Handling and Storage   39
Performance   41

4   Contact Cement   45
Chemical Basis and Behavior   45
Formulation   46
Handling and Storage   48
Performance   50

5   Hot-Melt Adhesives   53
Chemical Basis and Behavior   53
Formulation   54
Handling and Storage   57
Performance   57

6 Urea and Resorcinol Resin Glues   **60**

Chemical Basis and Behavior   60
Formulation   61
Handling and Storage   63
Performance   68

7 Epoxy   **74**

Chemical Basis and Behavior   74
Formulation   74
Handling and Storage   77
Performance   82

8 Polyurethane Glue   **85**

Chemical Basis and Behavior   85
Formulation   86
Handling and Storage   87
Performance   88

9 Cyanoacrylate Glue   **91**

Chemical Basis and Behavior   91
Formulation   92
Handling and Storage   96
Performance   99

10 Using Glue Successfully   **101**

Preparing Your Shop   101
Preparing Your Materials   105
Gluing Up   107
Conditioning   114

11 **Edge and Face Gluing** **116**

Solid Stock 116
Bent Laminations 121
Sheet Goods 124
Conditioning Edge and Face Joints 126

12 **Veneering** **127**

Choosing Substrates 127
Preparing Substrates 129
Preparing Veneer 130
Gluing Substrates 132
Gluing Veneer 134
Conditioning 141

13 **Assembly Gluing** **142**

Stresses That Affect Assemblies 142
Joint Design 144
Gluing Surfaces 144
Making Joints 144
Clean-up 147
Strengthening Assembly Joints 148
Adhesives To Use 150
Conditioning 151

14 **Decorative and Specialty Gluing** **152**

Solid-Wood Decorations 152
Veneer Decorations 153
Non-Wood Materials 158
Repairs 160

Glossary 164
Index 168

Introduction

When I began working wood professionally 20 years ago, glue was the farthest thing from my mind. Wood, tools, and machinery captured all my attention and my fancy. As I used glue I never paid much attention to the stuff because it didn't seem as though gluing required a whole lot of thought. It took a few years for me finally to figure out that my indifference to glue was restricting my progress as a woodworker. Once I started paying attention to adhesives, I was amazed at how much there was to learn about them that was worth knowing and at the difference that this knowledge made in my work.

With that in mind, I was also surprised that someone hadn't already written a book like this one to help me and other woodworkers learn more about glue. I would have been happy to have had such a book in my shop years ago so that I could have kept working at the bench instead of poring over reams of documents and making countless phone calls to learn more about glue. Nevertheless, I have greatly enjoyed writing this book because it has put me in touch with scores of knowledgeable and helpful people who understand how important adhesives are and who are as interested as I am in seeing woodworkers learn more about them.

There are many different adhesives available for use with a variety of materials in all sorts of situations—too many, in fact, to cover in one book. This book focuses on the glues that are used in small shops to build cabinets, furniture, and other projects out of wood, and to make repairs. The first section of the book deals with the adhesives themselves. It answers the question: "What should I know about the glue I'm using?" The second section of the book looks at various woodworking operations, such as veneering, from a glue standpoint. It addresses the question: "What glue can I choose for a particular job?" One or the other of these approaches should answer most questions that you may have about adhesives.

As you broaden your gluing knowledge and experience, you may wonder if glue and adhesive mean the same thing. There is a formal distinction, but it isn't widely observed. Strictly speaking, glues are derived from a natural source, such as plant starch (rice glue) or animal protein (hide glue). Adhesives are the product of synthetic polymer chemistry—hot melt or epoxy, for example. In this book, I use the terms glue and adhesive interchangeably, which is common practice.

Having a good fundamental understanding of adhesives will not only improve your work, but will also help you avoid problems, and not many woodworking problems are worse than gluing problems. But more than that, understanding adhesives gives you options that you otherwise wouldn't have. You can plan every aspect of a woodworking project from design to finishing based on the adhesives you choose. You can directly influence how easy or hard a project will be to build and how well it will hold up over time. The benefits aren't just technical ones; knowing how to use glue wisely will also help you be more creative and resourceful. As a woodworker, I want to have those sorts of advantages on my side. Wood is not always the easiest material in the world to control. In my shop, I'll take all the help I can get.

1

Understanding Glues and Adhesives

The most fundamental rule of the glue kingdom is that there is no such thing as a single all-purpose "miracle" adhesive, and there probably never will be. Different glues have different advantages and disadvantages, and some are better suited to certain tasks than others. The greater the number of gluing tasks you're doing, the greater the variety of glues you're likely to need. Becoming familiar with a wide range of adhesives will allow you to approach each stage of the woodworking process—designing, making, and finishing—from the standpoint of using the right glue in the best way possible. Once you do this, both the way you work and the results you get will be greatly enhanced.

This chapter will help you organize your thoughts about glue in general so that as we look at various types of glues in the eight chapters that follow you'll have a framework for learning more about each one. Once you begin to

compare the properties of different types of glues, you'll get a better idea of what to look for in a specific product (see the sidebar on p. 8) and how to relate it to your own work.

Even though the characteristics of adhesives are overlapping and interrelated, I've grouped them into four categories to help you think about them in an orderly fashion:

• Chemical basis and behavior
• Formulation
• Handling and storage
• Performance

Chemical Basis and Behavior

As woodworkers, we probably could get along without knowing much about how adhesives actually form bonds. I'm convinced, though, that I'm a better woodworker for having learned more than I once thought I needed to know about this. Because most of the adhesives we use are buried in our work where we can't see or touch them, it's helpful to have some insight into how they actually hold our work together.

When we use adhesives, we're not bonding wood to wood. We're bonding wood to glue. Wood and its by-products are very bondable materials. Like paper, they're fibrous, porous, and absorbent. The solid wood we work with is virtually unprocessed, having only been sawn and dried. Its cell structure is intact, with an orientation that is well suited to adhesion.

Wood is also a material whose defining essence is water. It gives up a large amount of water during the drying process, but also retains a fair amount of moisture even after being dried to well under 10% moisture content (MC). Once dry, wood also absorbs and releases moisture quite readily. These properties make it very compatible with many glues that are either water based or moisture curing.

The foundation of all adhesives is the polymer, a long chain of molecules that contains repeating structural units (monomers) linked together with strong "spines" of single-bonded carbon atoms. Polymers are commonly found in nature and can also be created in the lab by chemical synthesis. Wood itself is a monument to natural polymers. It is composed primarily of cellulose, the polymer found in all plant matter. Its cell walls are held together with another, more complex polymer, lignin. There are two types of adhesive polymers: homopolymers, which are composed of only one type of monomer, and copolymers, which are composed of more than one type of monomer (see the drawing at right).

A variety of glues helps ensure success with different kinds of gluing jobs. Buying small amounts of each glue keeps cost and waste to a minimum.

POLYMER STRUCTURE

Polymers are made from monomer molecules. If a polymer is composed of only one type of monomer, it's called a homopolymer. If it's made up of more than one type of monomer, it's called a copolymer.

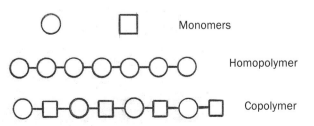

Learning More about Specific Adhesives

There are many different types and brands of glue on the market, and deciding which one to use for a given project can be a daunting task. It's hard to tell how these products differ from each other, just by reading the labels on the container. Fortunately, good information on glue is easy to come by. Here are four good primary sources of facts about glue:

• **Material Safety Data Sheets.** A Material Safety Data Sheet (MSDS) is a printout of the health and safety-related data for a given material, mainly those that have some sort of chemical content. There should be an MSDS available for every glue you use. They can be obtained from suppliers and manufacturers, usually on request.

• **Tech data sheets.** Many adhesive manufacturers print specification sheets for their industrial customers. Although not all the information on these tech data sheets is of interest to small-shop woodworkers, there are still a lot of helpful specifications that are worth having for reference.

• **Manufacturers and suppliers.** Chemists and product managers at adhesive manufacturing companies and suppliers who know their product line well can be very helpful. People who sell glue want glue users to be successful and usually provide good product support.

• **The Internet.** The amount of adhesive information available on-line is already extensive and constantly growing. It includes technical articles, product information, and the results of woodworkers' interaction, such as chat groups.

TYPES OF ADHESION

Adhesives bond wood in two ways: mechanical adhesion and specific adhesion. Mechanical adhesion is what happens when glue penetrates the surface of the wood and creates a slight physical connection by "grabbing the wood by its pores" as it cures and hardens. However, this mechanical connection is only a secondary form of adhesion.

Specific adhesion is the backbone of any adhesive bond, and it is based on intermolecular attractive forces. At first glance, polymer molecules don't appear to have a great deal of built-in potential for bonding with one another. They are stable and non-polar, which means that they don't carry the positive and negative electrostatic charges that are the basis for ionic bonding.

Polymer molecules do have a mechanism for bonding, though. They have the ability to induce each other to distort their structures temporarily and for a very brief time become dipoles, which are electrostatically attractive. At the atomic level, these attractive forces are not very strong, but at the molecular level, the force is greatly magnified, and it increases as the size and mass of the molecules involved increase. Polymer molecules are huge, with a high molecular weight and a large surface area. Because of their great size, they distort much more readily into temporary dipolarity than smaller molecules. Add this all together and factor in a glue joint's large surface area relative to its adhesive volume, and

Adhesives convert to solid substances in three different ways. PVA glue (left) cures through evaporation of water. Epoxy (center), a two-part adhesive, cures by means of a chemical reaction. Hot melt (right) cures through loss of heat.

you've got some attractive forces at work that are more than powerful enough to hold two pieces of wood together.

MEANS OF ADHESION

All adhesives undergo a fundamental change in order to form bonds. Each adhesive has its own means of converting from a liquid or a workable solid into a hardened cohesive solid. The conversion properties of the adhesives we use fall into three categories:

• **Evaporation.** Adhesives that contain water and/or other solvents cure to a solid as these fluids are transferred to the atmosphere and the glued work. For example, PVA glue cures by releasing its water, and flammable contact cement cures by releasing toluene and methyl ethyl ketone.

• **Chemical conversion.** Some adhesives contain components that will produce a chemical reaction when induced by one of several means. The chemical reaction converts the adhesive to a cured solid. A reaction can be triggered by moisture, as with polyurethane glue, by a catalyst, as with urea resin glue, or by blended formula components, as with epoxy.

• **Thermal conversion.** The performance of certain adhesives depends on the presence or absence of heat. Glues such as hide glue are fluid and workable only when heated. Adhesives such as epoxy require specific minimum temperatures to be effective and develop superior ultimate properties if heated while they cure. Glues that can or must be heated to promote curing are known

Conversion Properties of Adhesives

Evaporation	Chemical Conversion	Thermal Conversion
Wheat paste, rice glue		
Fish, hide, bone, rabbit-skin glues		Hide, bone, rabbit-skin glues
Casein	Casein	Casein
PVA, EVA, vinyl acrylic	PVA (Type II)	PVA, EVA, vinyl acrylic
Urea resin	Urea resin	Urea resin
Resorcinol	Resorcinol	Resorcinol
Contact cement	Polyurethane	Contact cement
	Epoxy	Epoxy
	Cyanoacrylate	Hot melt

as thermosetting adhesives, and those that soften and/or become fluid when heated are known as thermoplastic adhesives.

Many adhesives fall into more than one of these categories, with some listed in all three (see the chart above). Most adhesives are listed in the thermal conversion column because they behave and/or perform differently at a higher temperature, even if it's only a 20°F increase in the temperature of your shop.

CROSS-LINKING

When glue cures, or polymerizes, polymers become organized into a solid matrix. In some glues, long polymer chains are oriented into unconnected strands. In other glues, molecular bridges form between these strands, and they become connected. This phenomenon is known as cross-linking (see the drawing on the facing page), and it gives the cured bond layer much more cohesive strength because polymer molecules are physically connected to each other instead of just being attracted. Cross-linking usually takes place over the course of several days or weeks. Glues that cross-link are generally more durable and water resistant than those that don't.

Formulation

Like finishes, adhesives are carefully formulated. The ingredients in an adhesive formula and the proportions in which they are blended give the adhesive specific properties that are worth paying attention to. If you know something about an adhesive's formulation and its

CROSS-LINKING

Polymers can form bonds without being linked to each other, but when they cross-link, they form bonds that are stronger and more durable than the bonds formed by unlinked polymers.

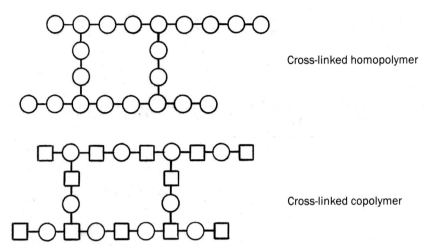

Cross-linked homopolymer

Cross-linked copolymer

physical properties, you'll know if the adhesive is appropriate for the job at hand and whether it will provide a high-quality, long-lasting bond after it cures. You will also know how safe it is to use.

The following list will give you an idea of what aspects of formulation and physical properties you should be aware of as you choose and use an adhesive. Refer to the chapters on specific glues to find out what the actual properties are for each.

• **Appearance.** Changes in a glue's appearance can help you monitor its quality and condition.

• **Color.** If the glue you want to use doesn't have the color you want for your work, you can alter the color of the glue or switch to another glue.

• **Grade.** Most commonly available woodworking glues are consumer-grade products, which are adequate

for many uses. If a consumer-grade glue doesn't suit your purposes, you can use an industrial-grade glue instead. Many industrial-grade adhesives are superior formulas that outperform consumer-grade products, and you can choose from among such glues on the basis of specific properties like solids content. However, some industrial-grade glues are economy formulas made primarily for low-cost, high-volume production woodworking, and these are inferior to retail consumer-grade glues.

• **Components.** Knowing an adhesive's components can tell you many things: how it functions chemically, whether it's hazardous, whether it contains filler or extenders, whether you can alter it, whether it's high quality, and whether it's been modified for enhanced performance.

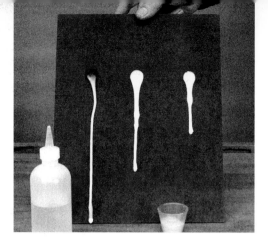

Altering viscosity. The glue in the middle is unaltered; the glue on the left was thinned with water; the glue on the right was thickened with wood flour from the cup. The same PVA glue was used for all three.

• **Viscosity.** If the viscosity of the glue you're using doesn't suit your work, you can thin or thicken the glue (see the photo above) or switch to another glue. Some glues are thixotropic, meaning that they have a variable viscosity. These glues are thick when at rest, but become thinner and more workable when they are manipulated.

• **Surface sensitivity.** The quality of the bond an adhesive delivers depends on the surface conditions of the materials being glued. It's important to know how the glues you use will bond under different surface conditions so you can choose the best glue for a specific job. If you are working with a challenging material, such as a dense and oily tropical hardwood, your choice of glue becomes more critical.

• **Moisture content.** Some adhesives contain more than 50% water, which can have a dramatic effect on the materials you're gluing. Glue with a high moisture content will also shrink a lot as it dries. You can alter the moisture content of a glue, but it's usually easier to use another glue that has a lower moisture content.

• **Percentage of solids.** With most types of glue, high-quality formulas contain a higher percentage of solids than low-quality formulas. If you're shopping for top quality, you can use solids-content percentages to compare different brands of the same type of glue.

• **Type of solids.** Not all glue solids have adhesive value. High-quality glues have a high content of adhesive solids (such as resins) and produce a top-quality bond. Non-adhesive solids (such as extender, filler, and pigments) don't contribute to the quality of the bond, and glues that contain a high proportion of these types of solids often produce an inferior bond.

Handling and Storage

Using glue involves more than just applying it to the work. You've also got to handle it, prepare it, and store it, and if you do those things well, you'll get the most out of the glue.

Many aspects of adhesive behavior affect storage and handling. The following list will give you an idea of what aspects to consider when choosing a glue and how those aspects will affect the way you'll need to handle and store your glues. You can then refer to the chapters on specific glues to find out what the actual handling and storage requirements are for each. A general rule to keep in mind is that glue should be protected from extremes of temperature, humidity, and light.

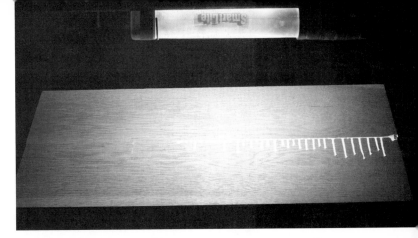

Altering glue with additives can be useful in various ways. Here, a PVA glue with a fluorescing agent added glows under a black light, making it easier to remove all traces of squeeze-out. The left half of the joint has already been scraped.

• **Shelf life.** All glues have a shelf life, and in many cases it's only a few months. You may be able to extend a glue's shelf life with proper storage. Some adhesives can be refrigerated to prolong shelf life. Keep track of shelf life, discard any glue whose age or condition you suspect, and don't buy glue that looks as if it has been sitting on a supplier's shelf for a long time.

• **Freeze/thaw stability.** Water-based glues freeze at 25°F to 30°F, and many of these glues can be ruined by freezing. Other water-based glues are freeze/thaw stable, which means that they will withstand one or more freeze/thaw cycles without spoiling. I don't like having to keep track of my glues' freeze/thaw stability, so I simply keep it from freezing at all times.

• **Preparation.** Woodworkers don't like to mix glue because it takes time and effort, but ready-to-use glues aren't completely trouble free either. They can degrade or polymerize in the container. They also generally don't deliver as high a level of performance as glues that require mixing. Mixing glue is worth the trouble, and it's not as much trouble as you might think. You can measure mix proportions by weight or by volume. Weight measurements are more accurate, but volume measurements are more convenient.

• **Alteration.** Many glues can be changed to suit different requirements. Mix ratios can be varied, viscosity can be altered, extenders or fillers can be added, and reaction rates can be accelerated or slowed. Plasticizers or hardeners can be added, the color can be changed, and different glues can be mixed together.

• **Pot life.** Pot life is the length of time that the glue will remain workable in a mix pot once the resin and hardener are combined. You can manipulate pot life through mix ratio, choice of mix components, or choice of additives. You can also control pot life by heating or refrigerating the glue mix and by using mix pots of different shapes and sizes.

• **Clean-up.** For health and safety reasons, your hands and skin should always be your primary clean-up concern. Your next concern should be the work you have just glued. You'll need to decide when and how you're going to clean up the excess glue depending on what glue you used and how you used it. Your final concern should be your gluing equipment. When cleaning glue off of your body, your work, or your equipment, use the least toxic and least aggressive means possible, especially on your skin. Safe cleaning materials include waterless hand cleaner, soap and water, and vinegar. Toxic solvents such as acetone are effective but should be

Getting glue on your skin is often unavoidable. Always use the least toxic means possible to remove it, starting with water, hand cleaner, or vinegar.

Three generations of adhesives in one chair: The chair was made with hide glue (a good choice), repaired with PVA glue for the sake of convenience (a poor choice), and repaired once again with carvable epoxy (the only choice left).

used only as a last resort. Generate as little waste, such as glue-fouled rinse water, as possible.

• **Disposal.** Old, spoiled glue and excess freshly mixed glue both need to be disposed of as safely and responsibly as possible. That means you should be familiar with federal, state, and local regulations regarding proper disposal. Because of the waste, inconvenience, and hazards involved with the disposal of glue, it's wise to buy and mix conservative quantities of glue whenever possible.

• **Health and safety.** Synthetic adhesives are produced from many different chemical groups, and some of them can be noxious, volatile, flammable, or pathogenic. Even natural glues have additives. Health and safety considerations should enter into all aspects of adhesive use, including mixing and application, clean-up, exposure to curing adhesives, and hand-tooling or machining cured glue. The National Institute for Occupational Safety and Health (NIOSH) has studied these issues extensively. You can obtain lots of data on glue health and safety issues from NIOSH by telephone (800-356-4674) or online (http://www.cdc.gov/niosh).

Performance

You can learn a lot about glue by simply putting it to work, and much of what you find out is stuff you'll never learn from any factory rep or counter clerk. But if that's the only way you learn what glue is capable of, you tend to make your work suit what you know about glue instead of finding the best glue for the work you have in mind. Being familiar with the performance aspects (listed below) will help you decide what to look for in an adhesive so that you can compare different products and use the right one for the job every time.

Because adhesives have lots of performance-related aspects, I have divided them into two categories: application and assembly properties, and post-assembly properties. The categories are organized in rough sequential order to help you locate and review individual properties more quickly.

APPLICATION AND ASSEMBLY PROPERTIES

• **Spread.** When you're applying glue by hand in a small shop, it's easy to forget that some adhesives and some types of work require specific spread thicknesses. Industry controls spread thickness with glue-spreading machinery; your best bet is to develop a feel for applying the right amount of glue by hand.

• **Workability.** Some adhesives are harder to apply than others because they're thicker, tackier, or older. Glue with poor workability can be hard to spread evenly and makes big glue-ups much harder to manage. It's difficult to alter a glue's workability; choosing another glue is a much better option.

• **Working temperature.** Most glues have a minimum working temperature. When used below this temperature, they will perform poorly or produce bonds that fail. If you store glues below this temperature, they should be warmed up before use. If you have a cold shop, you'll also have to warm your work before gluing it up.

• **Initial tack.** A glue's initial tack is the speed and strength with which it allows glue-spread parts to grab each other once they are assembled. Some glues have high initial tack, and some have none at all. Be mindful of initial tack as you choose which glue to use for different jobs so that the tack will be a help, not a nuisance or a hindrance. Glues with high initial tack aren't necessarily fast acting. Fish glue, for example, has a high initial tack, but sets and cures slowly.

• **Assembly times.** The amount of time that a glue will remain workable after you apply it is described in terms of open and closed assembly times. Open time is how long the glue is workable between the time you apply it and the time you assemble the pieces, while closed time is how long the glue is workable between assembly and clamping. These times tell you how fast you have to move during a glue-up, and it helps to know what they are before you use a glue for the first time. You can extend the assembly times of many adhesives by raising or lowering the temperature of the glue itself and/or the work, by changing the mix ratio, or by changing the mix components (e.g., choosing a slower hardener).

• **Clamping pressure.** No adhesive's performance is improved by excessive clamping pressure; some glues actually produce a weaker bond when too much clamping pressure is applied. If your work is well crafted, you shouldn't have to apply any more pressure than is needed to bring the work snugly together.

• **Setting speed.** Clamp time is governed by how quickly the glue you used sets up. When glue has set, its

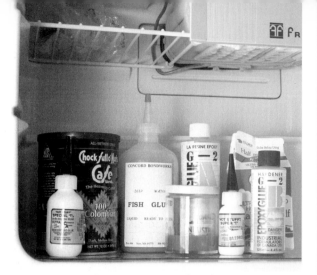

Refrigeration can help keep glue fresh beyond its stated shelf life. It's also a great way to slow down the reaction rate of some glues to allow more working time.

The boards in this panel were glued and clamped with PVA until the squeeze-out seemed set. The bond hadn't yet cured to a reasonable strength, though, and the edge joint was easily popped apart by hand.

bond is not at full strength, but you can unclamp and lightly handle the work. The setting speed of most glues can be adjusted by raising or lowering the temperature of the work.

• **Clamp time.** Because no shop ever has enough clamps, it pays to know the required clamp times of the different glues you use so that your clamps won't be tied up needlessly for hours. Clamp times can be greatly reduced by changing the temperature, and it's often very helpful to warm or chill your work as needed.

• **Penetration.** Glue must wet and penetrate the joint surfaces of wood and other porous materials to produce the best possible bonds. But penetration has to be controlled, or glue can bleed through the pores of veneer or be overabsorbed by porous materials, which can create a starved bond layer. Penetration can be controlled by choice of adhesive, thickness of spread, and clamp pressure.

• **Glue-line thickness.** All adhesives have their own optimal glue-line thickness. If the line is too thick, the bond won't endure stresses well, and if it's too thin, the joint will be starved of adhesive solids. The best method for guaranteeing decent glue lines is to craft good joints and not to overclamp them.

• **Curing requirements.** Some industrial adhesives may require curing conditions that are impractical in the average shop, such as temperatures in excess of 160°F. Glues designed for "cold" use will cure properly in average small-shop conditions. Some cold-use glues will work better and faster if cured with some added warmth or moisture.

Adhesive Ratings

Up until a few years ago, glue manufacturers had never given woodworkers much reason to think about adhesive ratings. Then Type II PVA glues were introduced, and woodworkers wanted to know what a Type II rating meant.

Essentially, adhesives are unregulated, but most glue produced in the U.S. is of high quality because there is a network of voluntary ratings standards in place. These standards have been developed by various organizations, such as the federal government (which rates adhesives for its own contracts), the military, industry groups, and independent standards organizations such as the American Society for Testing and Materials (ASTM) and the American National Standards Institute (ANSI).

Currently, adhesives are rated according to what kinds of bonds they produce. There are three types of rated bonds—Type I

(waterproof), Type II (water resistant), and Type III (non-water resistant). Bonds are rated based on a series of exposure, immersion, and boiling tests. This rating system, which was developed many years ago, is now considered imprecise and outdated because it doesn't allow for a gradual scale of relative performance. A glue that barely qualifies for a Type II rating can be advertised as being the equal of a glue that has the highest conformance to the Type II standard while in fact, there may be a big difference in performance between the two glues.

To make matters worse, these bond-type ratings have become generic. Many different organizations have adopted the rating system, and they often change the standards guidelines to suit their own needs. Thus, a plywood manufacturing association's concept of a Type II bond could be very different from the government's concept of such a bond. When a glue is advertised

Water-resistant PVAs such as these were introduced to woodworkers as Type II adhesives—an often confusing designation that is based on technical standards but is used more as a marketing tool.

as a Type II adhesive, it's often hard to tell whose idea of Type II the advertiser is talking about. If you want to use a Type II glue without having to sift through the confusion, your best bet is to buy the highest-quality glue you can find, test it in your shop to make sure it does what you want it to do, and call the manufacturer for technical support if you have specific questions.

Are Adhesives Stronger than Wood?

Most woodworkers have heard the classic sales line: "This glue is stronger than the wood itself." Though that claim sounds impressive and exclusive, it actually applies to all the glues in this book except plant glues, hot melt, and contact cement.

The strength of an adhesive is gauged by comparing it to the cohesive strength of solid wood in controlled tests. Maple is the standard benchmark species used in adhesive testing because it's strong, dense, and difficult to glue well. The cured strength of most woodworking glues easily exceeds maple's average shear strength of 1800 psi, so when test samples are broken apart, it's the wood that usually fails, not the glue line. Because of this, it doesn't make sense to choose glues on the basis of strength alone. Instead, choose glues on the basis of other properties as well, such as assembly times or heat resistance.

GAP FILLING

- Stile
- Thick glue layers fill gaps between tenon cheeks and mortise walls.
- End grain of tenon
- Rail

The rail tenon is too thin to fit properly in the stile mortise, and glue has been used to fill the gaps. Most adhesives bond best in thin layers and do not provide good structural integrity when used in large, gap-filling volumes.

POST-ASSEMBLY PROPERTIES

• **Cure period.** Woodworkers pay much more attention to a glue's setting speed than its cure period, but glue bonds can take a day or more to cure to full strength. Some glues also shrink and lose moisture as they cure. Glued assemblies shouldn't be stressed or heavily worked until the glue has completely cured.

• **Structural qualities.** Adhesives are formulated to develop various characteristics as they cure that are important to the structural integrity of an assembly. These include compression strength, tensile strength, torsion strength, shear strength, shock resistance, and rigidity. Different glues exhibit different degrees of these qualities, and comparing the differences will help you choose the right glue for a job. Most of the time you can get along fine simply knowing whether a glue is rigid or flexible, brittle or shock resistant, and so on. If you're designing and constructing something that has critical structural and/or engineering requirements, though, such as a boat hull or an airplane propeller, you must make sure that your glue of choice specifically meets those requirements.

The Permanent Glue Joint: Fact or Myth?

Many woodworkers are certain that their best glue joints will never come apart. Unfortunately, there is no such thing as a truly permanent glue joint, even though some glues are known as "permanent" glues because their strength exceeds that of wood (see the sidebar on the facing page). All glue joints fail eventually, even superbly crafted ones bonded with the best modern synthetic adhesives. There are several reasons for this. One is that people abuse glued work, leaving it in wet basements or next to heating vents. Another is that cured adhesives degrade with age—they can dry out, become brittle, soften or weaken, even without suffering abuse. Finally, glue joints are constantly subjected to the forces created by seasonal wood movement. None of this is cause for alarm, and your projects probably aren't going to fall apart anytime soon. Glue joints can last for centuries, and have, even in conditions that are less than ideal.

• **Gap-filling ability.** In an ideal world, no one would need gap-filling glue. In the real world, however, woodworkers have always sought out glues that would fill structural and cosmetic gaps well. Many adhesives are advertised as being gap filling, but very few of them actually do a decent job of it.

• **Cured working qualities.** When glued work gets processed after assembly, so does the glue itself. Using glue that has good cured working qualities can make a big difference in how long a project takes and what it looks like when it's done. These qualities include machinability (how much and how fast glue will dull blades and cutters), sandability (whether glue abrades easily and whether it will load up sandpaper), and stainability (whether glue will accept or resist stains, dyes, and other finishing materials).

• **Endurance qualities.** If you want your work to last a long time, the glue you use has to be able to withstand

Some glues take stain better than others. Hide glue (left) accepts water stain readily, but cyanoacrylate glue (right) resists the stain completely.

lots of destructive forces. Consider using glues based on their resistance to: moisture, heat, light, weather (a combination of those three), solvents, and organisms (insects, bacteria, fungi, etc.).

2

Natural Glues

When I began working wood professionally, the only glues I used were synthetic. I wasn't aware of that because I never gave the type of glue I used a moment of thought. If someone had asked me if the bottle of white glue on my bench was natural, I probably would have said yes because I grew up with the stuff, and using it seemed completely natural to me. I had no idea that a world of truly natural glues—the adhesives that come from plant or animal sources—lay beyond my limited horizon.

Natural glues have been in use for at least 4,000 years and were the woodworking adhesives of choice until 50 years ago, when synthetic glues began to dominate the market. Nowadays, plant glues are largely ignored, and animal glues are largely misunderstood. Before I ever fired up a pot of hide glue, I had visions of it taking hours to make

and days to dry, coming apart at the first hint of moisture, and stinking up the shop. However, since the first time I used it many years ago, I haven't had any of those problems. Like all natural glues, hide glue is a useful, sophisticated adhesive.

Chemical Basis and Behavior

Natural polymer molecules are the foundation of plant and animal glues. Plant glues are based primarily on starches, and animal glues are based on collagen, the protein found in skin and bone. Protein polymers are larger than starch polymers and form stronger, more durable bonds due to their greater mass, which is why animal glues are used far more extensively than plant glues in woodworking.

As a group, natural glues remain as useful today as ever for a wide variety of jobs. Most are made from dry ingredients and water, ranging from simple rice and wheat pastes to refined industrial-grade formulas such as casein glue.

Protein and starch share some important characteristics. Neither is soluble in water, but both can be hydrolyzed, or broken down into particles when combined with water. With the addition of heat, these particles disperse into a stable colloid, which is what glue is, and become readily cohesive and adhesive. These properties give natural glues their characteristic tack and allow them to bond a variety of materials, making them highly versatile.

Natural glues harden as the moisture they contain evaporates and is absorbed by the glued work and the atmosphere. Hot hide glue also hardens as it cools. Since most natural glues don't undergo a chemical conversion, they can be reactivated with water (and heat if needed) after they harden, and are known as reversible glues (see the sidebar on p. 22). The exception is casein glue, which cures by both chemical conversion and evaporation and can't be reversed after it hardens.

This chair, made by the author, was assembled entirely with animal glues. Fish glue was used for secondary work, such as the back-slat mortise and tenon joints, and hot hide glue was used for all primary joints.

Reversibility

The ability of natural glues to be reactivated by water is one of the main characteristics that distinguishes them from synthetic adhesives. Some woodworkers think that this reversibility is a shortcoming and that joints bonded with reversible glues are inherently inferior to those bonded with nonreversible glues, but that's not true—reversibility is actually an advantage. Reversible glues are more versatile than nonreversible glues because you can redo your work if you make a mistake. They also make future repairs much easier, which is important if you want your work to last for generations. The reason that many masterfully crafted antiques have survived for centuries isn't because their hide-glued joints have never come apart, but because their joints were easy to reglue when they did come apart.

Plant-Based Glues

The plant-based glues most commonly used by woodworkers today are wheat paste and rice glue. Woodworkers who use plant glues either make their own from scratch or buy a refined dry formula and mix it with water. When you start from scratch (with uncooked flour or rice), you're in control of how the glue mixes and works, and you can experiment with additives to develop your own favorite recipes. Most woodworkers today, though, prefer to use the refined dry formulas.

FORMULATION
Refined powdered wheat and rice pastes are generally preferred because they are made from varieties of natural grains that are best suited for adhesive uses, they mix easily with water and produce consistent batches, and they save time, even though they still have to be cooked after being mixed. As an alternative, you can use one of the newer "pre-cooked" paste formulas, which don't require cooking before use.

Another recent improvement in some paste formulas is the addition of a chemical preservative, such as thymol or phenyl phenol, which prevents the growth of bacteria and fungi. When such growth occurs, it discolors the glued materials and breaks down the bond layer. If you buy a paste formula that doesn't include a preservative, you can add one as you make the paste.

HANDLING AND STORAGE
Like plain flour or rice, refined powdered paste formulas should be stored in a dry location so they will always flow freely and mix readily. When you mix paste, use the best-quality water possible (see pp. 103-104). Also, use sparing amounts of water to mix the paste; if it's too wet when applied, it will wrinkle paper and parchment or may deform cloth and leather.

In this screen by Carl Swensson, the paper is glued to the wooden framework with wheat paste, which dries clear. Any excess can be pared away with a sharp hand tool after it dries. (Photo by Sloan Howard.)

Cooking the paste after mixing can be tedious and time-consuming. To speed up the process, you can use a microwave oven. It's a good idea to make wheat or rice paste only as needed because wet batches usually don't keep for more than a few days. However, batches made from some refined formulas may keep for a few weeks. You can refrigerate mixed paste to help keep it from spoiling.

Wheat and rice pastes are completely safe and nontoxic and clean up easily with water. Cured squeeze-out can be cleaned up readily with hand tools as well as with water, but it tends to clog sandpaper. Don't dump excess paste down the drain, though. Add it to your shop trash instead.

PERFORMANCE

Plant-based glues have long open and closed assembly times and a long cure time, which makes them forgiving and easy to work with. Because they also have good initial tack, some types of work can be assembled, pressed down with a roller or burnishing stick, and left to dry without clamping. These glues dry to a clear, flexible film that will take aniline dyes and won't inhibit wood finishes. Because they are readily reversible with water (see the sidebar on the facing page), plant-based glues don't have good moisture resistance. They are fairly heat-resistant if kept dry, though.

Fish Glue

Fish glue is what the name says it is. It's processed from the skins of various fish species in different parts of the world. In North America, fish glue is made from cod and haddock that are caught commercially in the North Atlantic. The glue is used in various grades by industry for a wide range of woodworking and non-woodworking applications.

Fish glue is not widely used today, but it's a highly versatile glue with many useful properties. Note the date on the tall bottle; if kept refrigerated, fish glue has a long shelf life.

FORMULATION

Fish skins are washed, then cooked in vats of water with either a mild acid or base added to hydrolyze the protein. Once all the protein has been extracted and is in solution, the liquid is put in an evaporator to increase the concentration of proteinaceous solids from about 5% to 45%. The finished glue includes a small amount of stabilizer (titanium dioxide) and a bactericide, which is added to prolong shelf life, as well as a mild-smelling aromatic additive (which is not unpleasant).

Fish glue is the handiest natural glue available. It's a one-part, ready-to-use glue that gels at about 40°F, so it is quite fluid at room temperature and will run

and drip readily if heavily applied. Since wood is not often glued up at temperatures below 40°F, fish glue hardens primarily through the evaporation of its water content.

Fish glue is the natural glue of choice for multi-material work, such as metal-to-wood bonding. It adheres to a wide range of non-wood materials, but the quality of the bonds it produces varies with the material. For sound, long-lasting bonds, at least one of the two materials being bonded with fish glue should be porous.

HANDLING AND STORAGE

If stored correctly, fish glue has a long shelf life. I've had glue that kept well for eight years before I finally used it up. I date bottles and keep them refrigerated. The added stabilizer may settle out and collect at the bottom of the container as the glue ages, but the glue can be stirred to remix the stabilizer or be used as is. This glue is also completely freeze/thaw stable.

Fish glue penetrates well and should be used sparingly or be thickened before being applied to thin, porous material so that it won't bleed through to the surface. It can be thickened with inert fillers such as colloidal silica to minimize bleed-through. However, thickening fish glue won't make it a good gap-filling adhesive. It shrinks too much as it dries, and it's too brittle to withstand the stresses of holding gapped joints together once it's cured. When thinned with water, though, it makes an excellent glue size or sealer.

Despite its brittleness, fish glue produces a bond that's very rigid and heat resistant, which makes it easy to sand. You can make the bond more flexible by adding small amounts of glycerine as a plasticizer. Use the glycerine sparingly—adding too much will weaken the glue's performance.

Fish glue can be tinted with colorants such as ink or aniline dye. It can also be added to other water-based glues such as hot hide glue or PVA to alter their performance properties. Adding fish glue can extend an adhesive's assembly times and increase its degree of tack, for example.

Though fish glue is normally easy to reverse with water, it can be made water resistant with chemical additives such as glyoxal or formaldehyde. Of the two, glyoxal is a better choice because formaldehyde is hazardous. With or without additives, fish glue is insoluble in organic solvents, such as toluol or lacquer thinner, once it's cured.

Fish glue cleans up easily with water even after it's completely cured. I often let the squeeze-out dry on assembled work, then clean it off with wet rags after the clamps come off. You have to clean surfaces thoroughly, though, to remove all traces of glue. You might rinse an area until it looks clean but find that it still feels slightly tacky. At that point, you can either rinse again with fresh water or let the area dry and then sand the surface. Any residual trace amounts left on the surface of bare wood will not interfere with finishing.

Disposal of fish glue is rarely an issue; I usually use it up before it goes bad. Spoiled fish glue has a separated appearance and performs very poorly. When it gets to that stage, I add it to the manure pile for the garden. The glue is completely safe to handle and nontoxic to work with.

PERFORMANCE

Fish glue has a high initial tack, which can make workability an issue. Brushing and rolling the glue can overwork it, causing it to form long, sticky strands. If this happens, slow down your application rate. The strands will blend back into the glue before it starts to set. When gluing cloth, leather, and other pliant materials, this high degree of tack should hold the assembled work together without clamps until the glue hardens.

Open assembly time is very generous—up to 1½ hours. Closed assembly time can be hard to judge. Due to its tackiness, fish-glued joints grab quickly and can seem like they're setting up right after you assemble them. But the glue actually sets slowly, and parts can be manipulated after assembly, if necessary. Move parts only as needed to avoid overworking the glue.

With its slow setting speed, fish glue usually requires hours of clamp time. Setting depends on conditions and the porosity of the wood or other material being glued. Only moderate clamping pressure is needed. Check the squeeze-out to make sure the glue has hardened sufficiently before taking the clamps off.

Hide Glue

Of all the materials I've used in my years as a woodworker, there are few that I would say are magical, or nearly so. Hot hide glue is one of them. Although synthetic glues can do amazing things that hide glue will never be capable of, the reverse is also true, which is why hide glue probably won't ever become obsolete. This glue takes more effort to use than PVA or polyurethane glue, and you may not realize its value until you've worked with it for a while, but if you put some time and effort into using hide glue, the payoff is tremendous.

FORMULATION

Hide glue is made from hide trimmings, which are washed, then soaked in a lime solution to prepare the protein for hydrolysis. The hides are then given a succession of water and acid rinses before being heated in large vats of water to extract the protein. Hides are heated several times to hydrolyze the protein. The hot protein solution from each heating is concentrated in an evaporator, then gelled, dried, and granulated. Aromas, preservatives, defoamers, and other components are added during the process.

Contrary to myth, top-quality hide glue does not smell bad—it's the cheap stuff that stinks. Good hide glue usually has a light, clean odor, which is not objectionable. Glue made from chromed tannery hides may have a slightly stronger odor than glue made from stockyard trimmings, though. Adding a

Hide glue is available in standard and high-clarity forms. Standard opaque glue is heavier bodied and is used for most types of work. High-clarity glue is preferred for specialty bonding by makers of furniture and musical instruments.

drop of vanilla or wintergreen oil to the mixed glue will mask any smell.

Grade Hide glue is available in different forms and strengths. It is commonly sold in both pearl and granular form. Granular hide glue is the material of choice—it's far superior to pearl hide glue. All the pearl hide glue sold in the United States today is imported and is often manufactured to shoddy standards under crude conditions. Much of it is actually bone glue (see p. 31), which is an inferior adhesive to begin with.

Granular hide glue is available in two basic forms: standard and high clarity. Standard glue has more body when mixed and heated and is opaque. High-clarity hide glue has a more delicate and refined character and is nearly transparent. Both are superb—I use the standard glue for everyday work and the high-clarity glue for specialty work.

Granular hide glue is also sold in a range of gram strengths, ranging from 135 grams to 512 grams (see the sidebar on the facing page). The glues most commonly used for woodworking are in the middle of the range: 192 and 251. If you are buying glue granules that are simply packaged and sold as a generic product with no gram strength listed on the container, you can either call the supplier or switch glues if you want to make sure you know what gram strength you are using.

Surface sensitivity High-quality hide glue will bond many different woods well, including hard-to-glue tropical timbers. This is because it has good inherent tack and adhesion properties, and also because of the defoamers that are added during the manufacturing process, which lower its surface tension

Gram Strength

Hide glue is manufactured in different gram strengths, so if you work with hide glue regularly, being familiar with the gram-strength scale will help you choose the right glue for your work. Gram strength isn't a measure of the power of hide glue; it's a measure of its molecular weight. Glue with a high gram-strength rating has a high molecular weight, which basically means that it has a high viscosity and a high concentration of protein. All gram strengths of hide glue are more than strong enough for woodworking.

Unlike cordless-tool battery voltage or motor horsepower, more isn't better when it comes to gram strength. In fact, it's worse. High-test glue (with gram strength of, say, 512) is so thick and gels so fast that it is virtually unusable under ordinary shop conditions. The glues at the other end of the scale are much better suited to woodworking. Mid-test glues, such as 251, are used for rapid assembly work. Low-test glues, such as 135, are useful when longer assembly times are needed. The lower a glue's gram strength, the more working time it will allow before it gels.

and increase its ability to wet and penetrate dense, oily surfaces. Hide glue also bonds non-wood materials well, such as bone, ivory, tortoise shell, and metal, provided that their surfaces have been well-prepared (as discussed on pp. 158-159).

HANDLING AND STORAGE

Hide glue has an outstanding dry shelf life when stored away from light, heat, and humidity. Granules have lasted up to seven years in my shop without degrading.

Preparing hide glue is not a big deal. Glue granules and water are mixed together, left to sit for a few minutes, then heated to 140°F. The first time you mix hide glue, use the recommended proportions of glue and water. Put the granules in first, then pour the water slowly over them and note where the granules and water levels fall so that you'll be able to judge the measurements by eye for subsequent batches. You'll need to measure every time you use a new gram strength, though, because each one has a different mix proportion. When your first batch of a glue is heated and flowing, note its viscosity, and use that as a guideline for future batches.

I use a commercial glue pot to heat hide glue, but these devices are expensive. Cheaper devices, such as beverage warmers, baby-bottle warmers, and electric slow-cook pots, are reasonable alternatives. I bought the commercial pot because I didn't want to fuss over the glue, and the pot has paid for itself by freeing me from that. The glue pot keeps glue at a steady, optimal 140°F. Hotter temperatures can break down the molecular structure of the protein, which shortens the life of the glue and weakens it. Cooler temperatures will make the glue gel too fast as you use it.

If you use a commercial glue pot, don't mix glue directly in the removable metal liner. Instead, use the setup that's

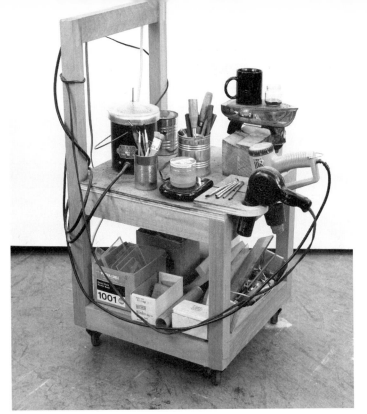

This impromptu hide-glue cart holds various heat sources: a glue pot, a cup warmer, and an inverted iron. It's also very helpful to have a hair dryer and a heat gun.

shown in the drawing on the facing page. With this method the glue never skins over, and it lasts longer as well. I make glue in small amounts, and a batch lasts me several days without degrading.

When I use a batch of glue for several days, it loses some of its moisture and thickens. I add water as needed to restore its proper consistency. If I thin the glue too much, I thicken it by leaving the lid of the pot off for a while until the glue skins over, then stir the skin back in. I also alter the viscosity slightly as needed for different tasks. Go easy if you do this: If the glue gets too thick, it will be hard to work with; if it gets too thin, you'll get a starved joint and a poor bond.

If I'm not going to use a batch of glue for a while, I cap the glass jar and refrigerate it. After a batch has been reheated several times, I make a new batch. I thin the old glue with water, and refrigerate it for use as a size or conditioner.

I prefer to clean up hide-glue squeeze-out while it's a workable gel because cleaning up is much more tedious after the glue hardens. To clean up cured hide glue you have to use both heat and water (I use rags, brushes, and a heat gun). I discard old hide glue by pouring it onto the manure pile for the garden. Hide glue is nontoxic and safe to handle, but always be careful when using heat with it so you don't burn yourself or damage your work.

PERFORMANCE

While hide glue isn't the most convenient glue to use, if you take the time to get accustomed to how it performs, you'll find that it's actually very handy stuff. Over the years, I've learned how to use hide glue's performance properties to my advantage, and now I couldn't imagine working wood without it.

Assembly time Even though hot hide glue flows smoothly, it doesn't have great workability or long assembly times because of its high initial tack and rapid gelling. To improve these characteristics, choose glue with a lower gram strength or raise the temperature of the glue, the shop, or the work. Of the temperature options, warming the work is the most efficient. I warm up the work with anything that's cheap, safe, and available, from sunlight to heat lamps. Fish glue can also be added to hot hide glue to lower hide glue's gel temperature and its setting speed, which will allow longer assembly times.

Another way to extend assembly time is to add a gel depressant, such as urea or table salt, to the glue. (Urea is available at pharmacies and garden centers.) Depressants retard the glue's gel rate and allow you more time to spread glue and assemble work. Add small amounts of depressant to begin with, and increase the amount as needed. Start by adding urea or salt to the granules at 2% to 5% by weight. If that doesn't slow down the gel rate enough, you can add up to 20% urea or 12% salt. When you add depressant, the glue will have a thinner consistency and the required clamp time will increase. You may need to ease up on the clamping pressure to avoid starving your joints. Gel-depressed glue degrades in the pot faster than regular glue, so make it only as you need it.

If you were to add 30% or more urea to hot hide glue, its gel rate would become so depressed that it would remain liquid at room temperature. A heavily retarded liquid hide glue is convenient and forgiving because you don't have to heat it, and it allows very long assembly times. But it also sets and cures much more slowly than hot hide glue because it doesn't gel, and it converts to a solid only through the evaporation of its water content. When you use a heavily retarded hide glue, you'll always have to clamp your work, and will have to leave the clamps on for extended periods of time. Also, the cured glue may soften in summer heat and humidity extremes.

As liquid hide glue ages, it begins to gel at room temperature. The remedy? Treat it like hot hide glue, and warm it in

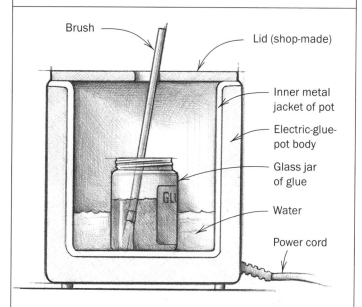

CUTAWAY VIEW OF A HOT GLUE POT

Brush

Lid (shop-made)

Inner metal jacket of pot

Electric-glue-pot body

Glass jar of glue

Water

Power cord

The two keys to improving hide glue use are to make the glue in a glass jar, which sits in a jacket of hot water, and to keep a lid on the pot, with a cutout for brushes.

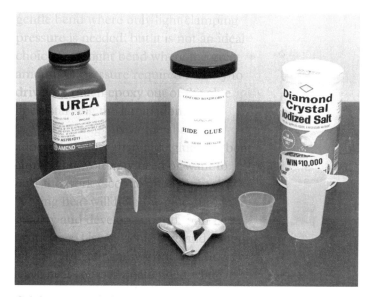

Gel depressants help retard the gelling of hide glue, which allows more working time. Don't add depressants routinely, but use them only as necessary because they affect the glue's overall performance.

Aging liquid hide glue, which begins to gel at room temperature, can be brought back to a usable state by warming its container in hot water. Here, an upside-down iron serves as the heat source.

Hot hide glue's natural tack will hold the knee block in place without clamps until the glue gels. As the glue cures, it will shrink, pulling the block tightly to the leg and rail.

some hot water (see the top photo at left). Once you've warmed it, you'll always have to warm it, but it can still be used in the squeeze bottle, which is very handy.

Clamp time With hide glue's high initial tack and rapid gel rate, most unstressed assemblies don't need to remain clamped for long periods. You can gauge the clamp time by the squeeze-out. When the gelled glue can be neatly peeled from the work, it can be unclamped a few minutes later. There are also many cases, such as with rubbed edge joints, glue blocks, and hammer veneering, where no clamping is needed. If the parts are well prepared, hand pressure will bring them together tightly enough to allow the glue to grab properly and hold them. As the glue gels and releases moisture, it shrinks a lot, and the shrinkage pulls glued workpieces more tightly together. Clamped or not, hide-glued joints need at least a day to cure to full strength.

Structural qualities Cured hide glue is very strong, with a tensile strength almost three times that of epoxy. Even though it can technically be reversed by heat and moisture, on a practical basis it's very resilient. It also cures to a rigid, creep-resistant glue line that has good sandability.

If you need a glue line that has some flexibility, you can plasticize hide glue by adding some glycerine to it, in amounts up to 10% of its granular weight. As with gel depressant, start with small amounts and work up from there. Use less water in the glue mixture if you add glycerine, and expect lower initial tack, a slower gel rate, and longer clamp times.

Cured working qualities From a finishing standpoint, hide glue is about the best glue you can use. You can tint it with ink or aniline dye before application if needed to color-match it to certain woods. Usually this isn't necessary because cured hide glue has a natural appearance in joint lines that blends in perfectly with many woods. Cured hide glue also takes water-based stains well and won't inhibit the application of finishes. In fact, hide glue is often used as a surface conditioner to prepare woods such as cherry and pine for staining and finishing.

Endurance qualities Though normally reversible with heat and water, cured hide glue can be made water resistant by adding either a metallic salt, such as aluminum sulfate, or formaldehyde (which is toxic and not recommended). Add the metallic salt sparingly as you make the glue—the proportion should not exceed 1% of the dry weight of the glue granules in the batch mix.

Rabbit-Skin Glue

Don't tell the kids, but the skins of rabbits make great glue. Rabbit-skin glue has traditionally been thought of as a delicate glue that's better suited for gilding and artwork than for woodworking. In fact, rabbit-skin glue can be used instead of hide glue for many woodworking tasks. On a practical basis, it's virtually the same thing. Sold in granular form, it's prepared and used just like hot hide glue, and has many similar characteristics. It's used extensively in Europe by furniture restorers for structural repairs and other demanding jobs.

The "pearl hide glue" shown here is actually bone glue. Like much of the bone glue available today, it is poorly made. Its dark color is a sign of its low quality. It cures to a brittle bond and has a smell you will never forget—hence the clothespin.

Bone Glue

Even though horns and hooves aren't used in animal glue making, bones are used to make glue in much the same way that hides are. Bone glue is no longer made in the United States, but lots of it is made in other countries. Its quality is very uneven. High-quality bone glue is used by European restorers for marquetry work. Low-quality bone glue is exported to the United States, where it is often sold as either pearl or granular hide glue. Poorly made bone glue is darker than hide glue and has a strong odor because it has been processed without preservatives and has putrefied.

Casein Glue

Casein glue is as close to a synthetic glue as natural glue gets. It cures by chemical reaction as well as by moisture loss, and is resistant to moisture once it cures. For these reasons, it was industry's ultimate adhesive weapon in the days before synthetic glues were developed. Today, casein glue is still used for many jobs

Casein glue should be mixed carefully, according to the manufacturer's instructions, to avoid lumping. A dust mask or respirator is a must. A cordless eggbeater does a fine job on small batches.

and whey. The curd is processed and refined, then blended with various inorganic alkaline compounds that allow the protein to disperse easily in water when the glue is mixed and cause the chemical reaction that occurs during the curing process. Preservatives are added to prevent bacterial and fungal growths from attacking the glue after it cures. Lower grades of casein glue also include soybean glue, which is added in various proportions as an extender. Although soybean glue is a good adhesive, it has little resistance to moisture once it cures. Adding it to casein glue lowers casein glue's water resistance.

HANDLING AND STORAGE

Casein glue is sold as a premixed powder that has to be stored well to preserve the stability of its chemical components. Shelf life is about 12 months if the powder is kept in a tightly sealed container away from heat, light, and moisture.

Airborne casein powder particles can cause upper-respiratory-tract irritation if inhaled, and the glue gives off a slightly irritating vapor when being mixed. Wear a respirator or a good disposable mask during measuring and mixing to help prevent both these problems. And wear gloves when gluing up because the alkalinity of the glue creates a soapy film on your skin that's hard to wash off and dries out your skin.

Casein glue has a specific mixing schedule that must be followed carefully. A power mixer is advisable to keep the glue from clumping during mixing. Once the water and powder are measured out, the powder is slowly added to the water as the water is stirred. The mix is stirred for several minutes, left to sit for 10 to 15 minutes, then

that other natural glues would never be used for, such as laminating interior structural timbers and doors, and it's an inexpensive, useful alternative to synthetic glues such as urea resin glue. It's also useful for gluing woods that are difficult to bond, such as tropical hardwoods, because it wets and penetrates dense, oily surfaces better than many synthetic glues.

FORMULATION

Casein glue is unique among animal glues in that the creatures that provide the raw material actually survive the glue-making process. Like milk paint, casein glue is made from milk curd, which is the proteinaceous solid that comes from separating milk into curd stirred again for several more minutes until the glue is smooth. If your batches turn out lumpy, try adding all the powder to half the amount of water, which will make a thick paste. Once this paste is smooth, add the rest of the water.

Casein glue can be cleaned from your skin and tools with water and ammonia or ammonia-based household cleaners while it's wet, but don't wash residual glue down the sink drain during clean-up. Excess mixed amounts can be discarded as a dried solid.

PERFORMANCE

Mixed casein glue is grainy but workable, and spreads adequately if the mix isn't too thick. Control the spread carefully to reduce squeeze-out and to make up for the glue's low initial tack, which allows glued parts to slide around once they're assembled and makes them hard to register for accurate clamping. The heavier the spread, the more fluid the closed assembly will be.

If the conditions in your shop aren't ideal, casein glue is a forgiving adhesive. It will cure in near-freezing temperatures, and its open assembly time lengthens as the temperature drops. Casein's open time in hot weather is at least 10 to 15 minutes. If that isn't long enough, the glue can be refrigerated right after mixing to extend its open time.

There's a flip side to this forgiving nature, though. At colder temperatures, casein-glued work can take days to cure and will produce weaker bonds. It's preferable to cure the glue at 65°F or above. If you warm assemblies while they're being clamped, try to keep them warm for a few days after they have been unclamped to allow the glue to cure to its optimal strength.

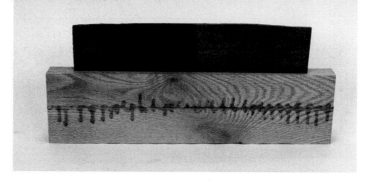

Casein glue stains many woods, such as the mahogany (top) and white oak (bottom) shown here. All excess cured glue has been removed from these samples; what remains are the chemical discolorations, which can penetrate well below the wood surface.

Once cured, a casein bond is tough, but not as strong as a urea or resorcinol resin bond. It's more shock-resistant than a urea resin bond, though, and is rigid enough for curved laminations. A cured casein bond is moisture resistant, but not waterproof. Cured casein glue is resistant to solvents, and is also very heat resistant and is nontoxic when cured. Because of its durable glue line and relatively low shrinkage during curing, casein glue is a better gap-filling adhesive than other water-based glues such as PVAs.

Cured casein glue is rough on tools—it will dull cutters, knives, and sawblades quickly. Another drawback is that it imparts a dark stain to certain high-tannin woods such as mahogany and oak (see the photo above). This trait makes clean-up of wet squeeze-out a bad option. Since cured casein glue is very sandable, the best strategy is to minimize the squeeze-out and then remove it with power sanders once it's cured. Once any glue-stained surfaces are cleaned up, the cured glue won't interfere with finishing, but it can produce noticeably dark glue lines in lighter woods.

3

PVA Glues

If PVA glues were ever outlawed, I think a lot of woodworkers would simply abandon ship. PVAs are so useful and so popular that they have become part of the essence of woodworking for many people. There are many types of PVAs. Each one of them has jobs that it excels at, and as a group they provide options and capabilities that I never would have thought possible a few years ago.

The term PVA refers to a family of glues that includes the adhesives commonly known as white and yellow glue. PVA stands for polyvinyl acetate, which is the main component used in formulating these glues. Once the first consumer-grade PVA—white glue—became popular, glue manufacturers introduced a second product—yellow glue. The new glue was superior in many respects to white glue and carried a fancy new name: aliphatic resin glue. When yellow glue was introduced, white glue was distinguished from it by being referred to as polyvinyl acetate glue. Consumers thought that white and yellow glue were fundamentally different adhesives, and that yellow glue was a revolutionary new product. In fact, yellow glue is simply a higher grade of white glue with some color added. It outperforms regular white glue because it's a superior formula to begin with.

The specific use of the phrase "aliphatic resin" to describe yellow glues is a masterful but misleading marketing ploy. The term "aliphatic resin" is a generic chemical description that could apply to all sorts of materials. By definition, all PVAs are aliphatic resin glues, no matter what color they are.

High-grade industrial PVAs, which are available from specialty suppliers, are superior to consumer-grade PVAs, but consumer-grade PVAs are more than adequate for many tasks. The main

PVA glues are a much bigger family of adhesives than many people realize, and the ones shown here are just a small sampling.

reason to use a high-grade industrial PVA is for specific performance properties, not just to get a better all-around glue.

Both consumer and industrial product lines include the newer Type II PVAs (see the sidebar on p. 17). These glues contain an acid salt catalyst, which induces a chemical reaction that causes the glue to cross-link and form more durable bonds as it cures. Because the catalyst is integrated into these one-part PVA formulas, they are known as catalyzed or precatalyzed glues. They look and act like regular PVAs. There are more specialized industrial Type II PVAs that are sold as two-part systems, with a separate catalyst that's added just before use. These are not nearly as convenient to use as the one-part Type II PVAs, but they do deliver a higher level of performance.

Yellow glue (right) is just white glue with a few extras. Besides the added yellow color, it has a higher-grade stabilizer, a higher solids content, and more added tackifier.

Chemical Basis and Behavior

Like natural glues, PVA glues are water based and are composed of solid particles that are not soluble in water. But where the adhesive solids in natural glues are derived from extracted and refined natural substances such as starches and proteins, the adhesive solids in PVAs and other synthetic glues are derived from resins that are manufactured by means of chemical synthesis. These manufactured solids can't be dispersed in water as readily as starches and proteins and must be integrated by other means instead. In the case of PVA glues, the synthetic solids are emulsified, a process that allows them to be smoothly and consistently suspended in water, the same way that fat particles are suspended in milk.

Though the composition of their adhesive solids is fundamentally different, PVAs and natural glues share the same general polymeric structure. Both types of adhesives are known as homopolymers because their polymers are composed of only one type of monomer, as discussed and shown on p. 7. As emulsions, PVAs in their liquid state appear stable and homogenous, but they're not very hardy, and they can deform or degrade with age or under poor storage conditions.

Ordinary PVAs cure by releasing moisture to both the glued work and the atmosphere and by drying to a solid bond layer composed primarily of PVA resins. The newer Type II PVAs cure in two ways: by releasing moisture and by a chemical reaction triggered by an added catalyst. The reaction enhances the glue's performance by causing the polymer molecules to cross-link, which creates stronger, more durable bonds.

PVA glue can be bought pretinted; it can also be tinted in the shop. Three common, easy-to-use colorants are ink, water-based aniline dye, and ground pigments. The first two will thin the glue, while the pigment will thicken it.

Formulation

The polyvinyl homopolymer base that PVA glue is built from is comparable to the plain white paint that's used as a tinting base at your local paint store. Glue formulators blend various components with the polyvinyl base according to the glue's use requirements, just as the paint store adds colors to the tinting base to make various shades of paint. In this way, formulators can give the glue a nearly limitless variety of physical characteristics.

COLOR

Most PVA formulas begin as a white glue, and are then tinted as needed with either dyes or pigments. These colorants are usually mixed in a dispersant first, then added to the glue. PVA glues can be custom tinted to order, or you can tint them yourself with water-based dyes, pigments, or inks.

When manufacturers add color to PVA glue, the amount can be excessive. Some heavily tinted yellow glues bleed color into the surface surrounding a joint and permanently tint the wood. Added color can be helpful in many ways, though. It tends to separate out of the emulsion when glue ages or is stressed by temperature extremes, so it's a good indicator of shelf life and shipping conditions.

Added color can also help you disguise glue lines. Pretinted "dark" PVAs that blend in with dark woods and stain colors are now available. These glues get a lot darker as they dry, though, so test to make sure that the dried color of the tinted glue you're using is right for your work. You can also use a dark glue with light woods to give glue lines a contrasting appearance, either as part of the design of a project or as a way to help make sure that all traces of excess glue are removed from wood surfaces before finishing.

VISCOSITY

PVA glues are sold in a range of viscosities. Consumer-grade white glue has a low viscosity and runs easily. Yellow glue is more viscous and clings better to joint surfaces. A lower-viscosity glue will coat joint surfaces better, and a higher-viscosity glue will produce less squeeze-out during glue-ups. You can thin a PVA if needed by adding a bit

PVAs come in a wide viscosity range. The thin glue on the left flows well inside joints such as dovetails once they're assembled, coating joint surfaces properly. The thixotropic glue on the right is thick when at rest and fluid when worked. It clings well to open joint surfaces instead of running or dripping off.

of water, but don't add more than 5% by weight.

Thixotropic PVAs appear to have a very high viscosity, but that's only when they're at rest. Once they are manipulated, they flow readily, then regain their resting thickness when they are left to stand again. All PVAs thicken with age, though, and eventually become unusable. A thickened older PVA can be thinned with water, but it's better to buy fresh glue.

MOISTURE CONTENT

When you choose a PVA glue, you should think of its moisture content as being the flip side of its solids content. A glue that has a solids content of 45% contains more water than anything else. During a glue-up, most of this water gets transferred to the work and can have a

EVA: a PVA Relative

Once I started exploring PVAs as a large family of adhesives, I also discovered other closely related glues. One such glue is ethylene vinyl acetate, or EVA. An EVA looks and behaves much like a PVA, but has different properties, some of which are very useful.

EVA glue was developed for use with decorative overlays that are difficult to glue, such as vinyl and metallic films. It has high initial tack and adheres well to a variety of materials, but dries to a soft, flexible film that you can jam your thumbnail into long after it has cured. It is reversible with water and allows long assembly times. Cured EVA can be easily reactivated with heat—in fact, it is used in many hot-melt formulas. Liquid EVA glue usually outperforms hot melt when both glues are used for the same application because it coats, wets, and penetrates gluing surfaces better than hot melt.

With its flexible glue line, an EVA can also be thought of as a PVA that will accommodate seasonal wood movement, although there are limits to this ability. EVA isn't rigid enough to use for assembling a whole piece of furniture, but it can be useful with traditional construction methods that ignore or violate wood-movement principles, such as breadboard ends on solid tabletops and molding applied to the solid sides of a cabinet. If you find that EVA is too soft for uses such as this, you can mix it with some PVA to make a glue that's harder, but that still allows some wood movement.

Along with vinyl acrylic glues, EVAs are now being used very successfully as assembly glues for melamine-faced cabinet parts. They adhere well to the show face of melamine panel overlays, so cabinet parts faced with these materials can be joined without the need for wood-to-wood contact.

These glues are PVA relatives: the glues at left and center are ethylene vinyl acetates (EVAs), and the glue on the right is a vinyl acrylic. The vinyl acrylic has better water resistance, and the EVAs have better freeze/thaw stability.

great effect on it. It also may take several days for the work to rid itself of the added moisture. Try to give moisture as much time as possible to evacuate, even though you can safely machine most PVA-glued assemblies several hours after unclamping them.

PERCENTAGE OF SOLIDS

Regular PVA formulas for woodworking contain between 45% to 55% solids. The higher the percentage of solids, the more substance the bond layer will have after the glue's moisture dissipates. However, simply having a high solids content doesn't guarantee that a glue will deliver a topnotch bond. To produce the highest-quality bonds, a PVA should contain mainly adhesive solids, such as resins, rather than nonadhesive solids such as fillers, extenders, and pigments. The total nonadhesive solids content in a high-quality PVA is usually less than 10% by weight.

Handling and Storage

Because PVA glues are simple and safe to use, it's easy to think of them as being maintenance-free adhesives that don't require any special attention. Actually, PVAs should be handled and stored as carefully as other adhesives so that you can make the best use of their wide range of properties and capabilities.

SHELF LIFE

Like other adhesives, PVAs will usually exceed their shelf lives if given proper storage. The claimed shelf life of a PVA, which is usually found on its tech data sheet (see the sidebar on p. 8), is based on storage at room temperature. The actual shelf life of the glue depends on

the temperature at which you store it. Shelf life increases as the storage temperature drops from 70°F and decreases as the temperature rises from 70°F. For example, a PVA stored at 90°F will have at most half of the shelf life it would have at 70°F.

Shelf life is also influenced by the formula of the glue. A rule of thumb is that the more exotic and high performance the formula, the shorter the shelf life. While an all-purpose PVA may have a six-month shelf life, a highly reactive cross-linking PVA might have only a three-month shelf life.

Because PVAs are multi-component emulsions, their contents can separate during their shelf life (see the drawing below). But separation doesn't necessarily ruin the glue. A PVA can

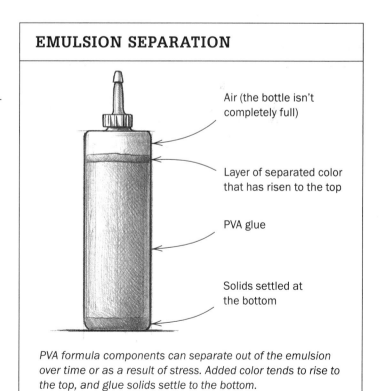

EMULSION SEPARATION

Air (the bottle isn't completely full)

Layer of separated color that has risen to the top

PVA glue

Solids settled at the bottom

PVA formula components can separate out of the emulsion over time or as a result of stress. Added color tends to rise to the top, and glue solids settle to the bottom.

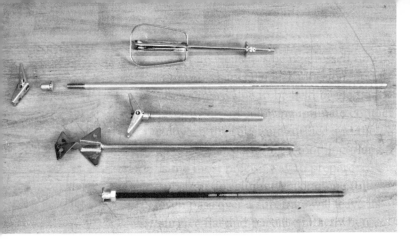

These stirrers are useful with PVAs as well as with other glues. Top to bottom: an electric beater blade; two shop-made stirrers fashioned from threaded rod, ground-down T-nuts, and toggle-bolt wings; a paint-mixing blade; and a shop-made stirrer with a threaded-rod shank and T-nut agitator for small glue bottles.

Toggle-bolt wings at the end of this shop-made stirrer are spring-loaded and collapse to enter the mouth of the gallon jug easily, then fan out for stirring.

often be stirred back into shape and used again. I use shop-made tools for stirring (see the photos at left). To determine whether stirring is necessary, probe the bottom of the container with a dowel (see the top photo on the facing page) to see if any solids have settled out of the emulsion and collected there. I check PVA glue for settled solids when I first buy it, and I find that stirring is often needed. In general, it's a good idea to stir PVA glue if it has stood undisturbed in its container for more than two weeks. The more time that passes between stirrings, the more solids you have to stir back into the emulsion.

FREEZE/THAW STABILITY

Because PVAs are water based, they will freeze in cold weather. Early PVAs were instantly ruined by freezing, but many of today's formulas can stand up to six freeze/thaw cycles before they degrade. Some precatalyzed cross-linking formulas are not freeze/thaw stable at all, however. You can tell at a glance when PVA has been ruined by freezing: The glue coagulates and separates, and can't be stirred back to a homogenous state (see the bottom photo on the facing page). In general, it's a good idea to keep any PVA from freezing, no matter how hardy its formula might be.

CLEAN-UP

Many woodworkers clean PVA from their hands and tools in the sink, but ideally, no glue should go down the drain. When I clean brushes and rollers, I wash them in a plastic tub full of warm soapy water. When the tools are clean, I generally leave the water in the tub overnight so the glue solids can settle to the bottom. I pour off the water on top and let the solids dry to an inert film, which I peel out and discard. You can pour old or excess PVA into shavings and let it cure to a solid, as shown in the photo on p. 114.

Cleaning excess PVA from glued work is a subject that has inspired a lot of debate over the years. There are three common clean-up strategies. The glue can be cleaned up with water while it's still wet, it can be removed when it has partially set, or it can be removed after it has hardened. However it's done, thorough clean-up of excess PVA on show surfaces is important because any glue left on the surface will inhibit finishing. PVA glue resists both stains and coatings and appears as an unsightly blotch wherever it remains on the work. No single clean-up method is best for all situations, so if you're comfortable with all the methods, it's a good idea to practice all three, choosing the right one for each glue-up. (For more on glue clean-up, see pp. 113-114.)

HEALTH AND SAFETY

PVAs are generally considered nonhazardous; small amounts that get on your skin are more of a nuisance than a danger. I wouldn't keep my hands in contact with PVA continuously, though, because it's acidic and can chap your skin. If the acidity of PVA glue causes you any concerns, you can buffer it to a neutral pH with calcium carbonate (chalk) or switch to an archival-grade pH-neutral PVA. Although PVAs typically don't pose much of a respiratory hazard, some formulas give off a noticeable acrid odor that isn't unpleasant but can be bothersome if breathed in large volumes for a long period. If you're spreading PVA over large surface areas, wear a respirator.

If there is a blob on the end of the stick, it means the glue solids have settled on the bottom, and the glue should be stirred before use.

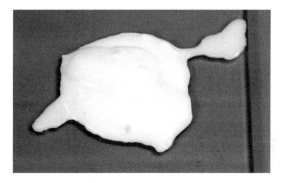

This PVA formula is not freeze/thaw stable. It has coagulated and separated and can't be stirred back into shape, as more stable PVAs can.

Performance

PVAs are generally workable adhesives. They flow smoothly and spread evenly without great effort. Lower-grade glues, such as consumer-grade white glue, are very workable because they have a lower viscosity, lower solids content, and less added tack than higher-grade formulas. Higher-grade glues tend to become

These samples were coated with the same glue. The sample at left was cured at room temperature; the sample at right was refrigerated at 38°F, which is below the glue's chalk temperature. Chalked PVA is very noticeable and bonds poorly, if at all.

stringy and overworked when rolled onto large surface areas. PVAs also become less workable as they get older.

WORKING TEMPERATURE

PVAs are fairly forgiving in terms of working temperature. They can be used in a hot or a cold shop, but they do have limits. PVA that's cold (below 60°F) doesn't flow as well as PVA at room temperature (70°F) and can be harder to apply. Lower-grade PVAs have poor heat resistance and produce soft, pliant bonds at temperatures of 100°F or more.

PVAs also have a specific low-temperature threshold, called a chalk temperature. If used at temperatures below this threshold, a PVA emulsion will lose its integrity as it starts to dry and will "chalk out" inside and outside a joint. Chalking is a frosty, powdery surface appearance that is very easy to spot (see the photo above). Chalked-out glue doesn't bond properly and dooms joints to certain failure. Chalk temperatures generally range from the mid-30s to low 40s.

In order to avoid chalking, the glue, the shop and the work all have to be above the chalk temperature. The work is usually the hardest of the three to warm up. If your shop temperature drops below freezing overnight and you warm up the shop the next morning, the work itself may not be warm enough for chalk-free gluing until several hours later.

INITIAL TACK

Basic PVA emulsions do not inherently have a high degree of initial tack. Tackifiers, such as refined pine resins, are added to PVAs to raise their level of tack. Some PVA formulas have more tack than others. A higher degree of tack is desirable for jobs such as face laminating, and a lower degree of tack is preferable for gluing joined assemblies such as door frames and table bases.

ASSEMBLY TIMES

One good reason to use a variety of PVAs is that different glues have different open and closed assembly times. Consumer-grade yellow glue has a rapid open assembly time, usually about 5 to 10 minutes at 70°F. You can buy glues that act faster or slower than that, and it's great to have glues with different assembly times on hand so you can choose the right glue for each job. Fast-acting PVA is great for rapid assembly work, such as adding glue blocks, and slow-acting PVA allows several extra minutes of open and closed time, which is very helpful when you're faced with a complicated, time-consuming glue-up.

It's also important to realize that the assembly times for all PVAs are variable. They can be influenced by lots of factors—shop temperature, humidity, airflow, the species and moisture content of the wood being glued, and the age of the glue and how thickly it is spread.

Open and closed times become longer at colder temperatures and faster at warmer temperatures. A PVA glue that has an open time of 20 minutes at 65°F will have an open time of 10 minutes at 85°F, and if there's a fan blowing 85°F air around the shop, the open time will drop to about 5 to 8 minutes. Closed assembly times are also influenced by how dense and dry the wood is. Hard maple that's been dried to 10% moisture content will allow a longer closed assembly time than poplar that's been dried to 6% because the softer, drier wood (poplar) will have absorbed more of the glue's moisture before the work is assembled.

PVA comes in different speeds (assembly times). Slow PVA (left) is good for tricky glue-ups. Regular PVA (center) is good for all-around work. Fast PVA (right) is good for quick assemblies and jigs.

SETTING SPEED

PVA glues generally set quickly, which helps keep clamp times short. As with assembly times, a PVA's setting speed depends not only on its formula, but also on the species, moisture content, and temperature of the materials being glued. Nonstressed joints in porous hardwoods that have been glued up in warm weather with an average consumer-grade PVA can usually be unclamped after less than 30 minutes of modest clamping pressure. You can test PVA squeeze-out with your thumbnail to determine how firmly the glue has set, so you'll know when you can unclamp the work.

CURING REQUIREMENTS

As a cold-use glue, regular PVA cures best at temperatures between 70°F and 90°F because the glue is thermoplastic and may not harden properly if exposed to higher temperatures. Catalyzed cross-linking glues will cure well at temperatures of 90°F to 100°F, and often they perform much better when heat cured. Some high-performance catalyzed PVAs actually require heat in order to cure at all.

CURE PERIOD

Even though PVA glue sets quickly, it still takes 24 hours or more (depending on its formula) to cure to full strength. Avoid stressing or vigorously handling and processing PVA-glued assemblies before they have cured, especially if they weren't glued up under ideal conditions.

STRUCTURAL QUALITIES

Cured PVA glue lines are strong and durable but are not very rigid. This lack of rigidity makes PVA more susceptible to creep than most other glues (for more on creep, see the sidebar on p. 106). Even though some high-performance industrial PVAs exhibit greatly enhanced rigidity, PVAs are generally not considered to be rigid-curing adhesives.

You can plane right through a cured PVA bead, which appears as the white border stripe on the shavings. The plane iron will dull eventually, but the edge won't get nicked.

Regular yellow glue and Type II glue were painted on a sample board and cured, then boiled in a pot. After a few minutes of boiling, the yellow glue on the left is easily scraped off, but the Type II glue on the right is much more resistant.

GAP-FILLING ABILITY

Some PVAs are currently being sold as gap-filling glues. These glues are filled with an inert additive, usually a wood flour, which gives them more body. But with their overall lack of rigidity, PVAs are not ideal gap-filling glues. They also shrink a lot as they dry, which doesn't help. If you need a glue with structural gap-filling ability, epoxy is a better choice.

CURED WORKING QUALITIES

Cured PVA glue lines are easy to machine and don't dull cutting edges nearly as quickly as glues such as urea resin or casein. I routinely push sharp hand tools through cured PVA glue lines without concern. The sandability of a PVA glue depends on how rigidly it cures and how heat resistant it is. Consumer-grade white glue cures to a fairly soft glue line and has low heat resistance, so it sands poorly and clogs sandpaper readily. Consumer-grade yellow glue is harder when cured and has better heat resistance, so it's much more sandable. Industrial-grade PVAs that cure to the most rigid glue lines and have the highest heat resistance are the most sandable glues.

ENDURANCE QUALITIES

PVAs exhibit a wide range of resistance to water. Despite all claims (see the sidebar on p. 17), none of the PVAs that woodworkers regularly use is completely waterproof; they are only water-resistant to varying degrees. Consumer-grade white glue is the least water-resistant PVA, and catalyzed cross-linking PVA is the most. If you want to push the limits of a glue you're using, request some technical literature so you can see what moisture levels the glue is capable of resisting.

4

Contact Cement

Before I went to work as an apprentice in a cabinet shop, I had no idea contact cement was such a widely used and important adhesive. My first year as a professional woodworker was a swirl of plastic laminate, contact cement, and routers as I learned firsthand how to crank out commercial casework.

In the years since, I've also learned how useful contact cement can be to have on hand no matter what kind of work you're doing. These days, I spend most of my shop time working at a bench with solid wood, veneer, and hand tools, and no longer use contact cement as a primary adhesive. Nevertheless, I still depend on it for some glue-ups and continue to find new uses for it. I can't imagine getting along without the stuff.

Chemical Basis and Behavior

Even though commercial and industrial shops depend heavily on contact cement, it is not considered to be a long-lasting (or "permanent") woodworking adhesive. The adhesive and cohesive properties of contact cement are not in balance: It will adhere to almost anything, but it has poor cohesion once it has cured. It never hardens the way other glues do. A film of cured contact cement is so soft and pliant that you can scrape it off a surface with your thumbnail.

Contact cements have low cohesion because they are based on synthetic rubber compounds, such as neoprene. Rubber can be produced in one of two states. In its basic state, it has lots of tack, but low strength and resilience. If rubber is vulcanized, it develops high strength

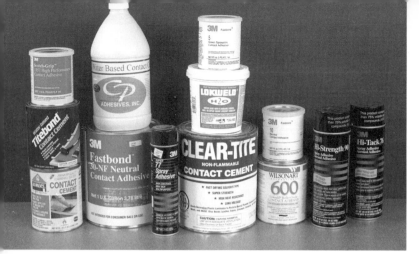

Contact cement comes in several forms, including flammable solvent based, non-flammable solvent based, water based, and aerosol.

and resilience but loses its tack. For adhesion purposes, formulators have to choose tack instead of strength.

Contact cement essentially consists of rubber particles dispersed in a liquid base. When contact cement is applied to a gluing surface, the liquid dissipates, leaving a dry film that will bond instantly to a similar cement film applied to a mating surface. The ultimate strength of a contact-cement bond depends on three things. The first is favorable shop conditions—the warmer and drier your shop, the better. Second, the mating cement films should be at the proper stage of readiness when they are placed in contact with each other. Films that are either wetter or drier than the application instructions recommend will not bond well. Finally, pressure must be applied to the work once it has been assembled.

Formulation

If you use contact cement only occasionally, you may be tempted to grab the most readily available brand

without giving it much thought. If you compare the specifications of various cements, though, you'll see that different formulas have different properties, and you can choose a cement that best suits your needs.

Most contact-cement formulas have similar types of neoprene rubber solids. The main difference between cement formulas is the type of liquid base those solids are dispersed in (see the sidebar on the facing page). There are three types of bases: flammable solvents, non-flammable solvents, and water.

Flammable solvents include toluene, acetone, and methyl ethyl ketone. The principal non-flammable solvents are 1,1,1 trichloroethane and methylene chloride. Water-based cements are, of course, carried primarily in water, but often contain small amounts of flammable solvents, such as toluene and methanol.

Contact-cement solids disperse readily in flammable and non-flammable solvents. In water-based formulas, the adhesive solids are suspended in an emulsion, like a PVA glue. The suspended particles are very fine and well dispersed, which gives water-based cements the same smooth, uniform consistency as solvent-based cements.

COLOR

Contact cement is now available in several colors, including red, green, and blue, as well as the original tan and yellow shades. Color is added so that the cement will present a contrast when being applied to neutral substrates such as flakeboard. As you apply tinted cement, the contrast helps you visually gauge your spread thickness and the thoroughness of your surface coverage.

Sorting Out the Solvents

No adhesive has ever caused more confusion and misunderstanding than contact cement has in the last few years. This is mainly due to repeated changes in regulations governing the solvents that are used to make some cements. People who use contact cements need accurate, helpful information regarding these glues, but many suppliers are confused and misinformed themselves. Here are a few basic details that will help you cut through the chaos and become a more informed consumer.

Twenty years ago, all woodworkers used flammable solvent-based contact cement, which works very well but is also very hazardous. Non-flammable solvent-based and water-based cements were gradually introduced as safer alternatives, and efforts to ban the flammable cements began.

Subsequently, the 1,1,1 trichloroethane solvent in non-flammable cement was found to be hazardous to human health and to the ozone layer. Plans to ban flammable cements were halted as regulators began working to ban non-flammable cements instead. Faced with these new bans, contact-cement manufacturers began making non-flammable cement with a different solvent, methylene chloride, which is a common ingredient in paint stripper (and which is also toxic). As they switched solvents, manufacturers were allowed to keep selling remaining stocks of the banned non-flammable cement. These stocks were enormous, which guaranteed that this cement would remain available long after it was banned and could be sold right alongside the new non-flammable cement.

Meanwhile, various regulations and tariffs have made it increasingly complex and expensive for suppliers to sell any solvent-based cement, flammable or non-flammable. Unsure of what they're allowed to sell or can afford to sell, some suppliers have stopped selling some or all of the solvent-based cements. Other suppliers simply are no longer sure what's in the cement that they are selling. To make matters worse, some suppliers have also begun referring to water-based contact cement as "non-flammable" cement. With further bans and regulations likely, this isn't the end of the story, but most woodworkers are willing to tolerate the confusion because contact cement is useful and convenient.

GRADE

Contact cement is sold in both consumer and industrial grades. As with other adhesives, industrial-grade contact cement generally outperforms consumer-grade material, and is available in a wider variety of formulas that are designed for specific tasks.

Contact cement is also sold in brush and spray grades. Brush-grade cement is formulated for brushing or rolling; it's thicker than spray-grade cement, which is applied with spray-finishing equipment. Consumer-grade cement is usually just sold in brush grade, while industrial cement is sold in both brush and spray grades. You can thin brush-grade cement for spraying, but keep in mind that brush and spray grades have about the same adhesive-solids content even though brush grade is thicker. If you thin brush-grade cement, it will have a lower solids content than spray-grade cement, and will deliver a weaker bond.

PERCENTAGE OF SOLIDS

Solvent-based adhesives contain from 18% to 25% solids by weight, and water-based formulas contain about 45% to 50% solids, with some new formulas approaching 60%. Try to use water-based cement with the highest solids content you can because it will have a lower moisture content. Water-based cement can add lots of moisture to the work, and this moisture can linger inside a closed assembly for weeks. The added moisture can swell substrates such as flakeboard and can warp plywood. These materials will also shrink as they dry out, which will cause joints and edges to shift and creep.

Handling and Storage

The claimed shelf life of contact cement is about one year, but the material will last for years if it's stored and handled well. Try to store cement at temperatures between 60°F and 80°F in tightly closed containers so the solvent or water won't evaporate. (The cement thickens and eventually becomes solid if the liquid evaporates.) Proper storage of solvents is also very important, because of flammability, health, and environmental issues.

Water-based contact cement is highly alkaline, so don't keep it in aluminum or galvanized zinc containers because it will react with those metals. Also, don't pump or spray any contact cement with equipment made from copper or copper alloys because those metals can break

Aerosol Contact Adhesives

Contact-type adhesives are available in aerosol form. These aerosols are close cousins to contact cements, but there are some fundamental differences. Their adhesive bases are usually styrene and acrylic compounds rather than neoprene, and most of them have a flammable solvent carrier, although some water-based aerosols are being introduced. Surfaces sprayed with aerosol adhesives have to be assembled while the cement is still wet. Bond strength depends on the two adhesive films wetting each other out as they make

contact. If the adhesive films dry out, they won't bond properly, but dry aerosol films can be reactivated with heat, solvents, or another spray coat of cement before assembly.

Aerosol contact adhesives are strong enough to be used for some structural applications. Here, high-density foam staves are being edge-glued into a coopered seat support for upholstery conservation work.

down the rubber adhesive base. Keep water-based cement from freezing because it has poor freeze/thaw stability.

If you're using spray-grade contact cement, it's best to dedicate an old or inexpensive spray gun to the task. Use a gun that has stainless steel or synthetic parts; do not use guns with components made from aluminum, steel, or copper alloys.

ALTERATION

Contact cement's viscosity fluctuates with temperature change. It gets thicker in a cold shop and more fluid in a warm shop. Cement can also thicken as part of its water or solvent base evaporates while it's being handled or stored. Try to develop a feel for a cement's optimal viscosity for various uses, and thin it as needed.

Water-based cement reduces easily with water, but if you add too much thinner to solvent-based cement, the adhesive solids will drop out of solution, and you'll end up with a clump of rubber in the bottom of the container.

Contact cement may separate somewhat in storage. If this happens, it can be easily remixed before use. Solvent-based cement that has dried out and solidified in storage can be reconstituted with solvents. However, dried-out water-based cement can't be reconstituted with water. You could try to reconstitute dried-out water-based cement with solvents, but it's better to discard it and start over with fresh material.

CLEAN-UP

If contact cement is used properly, there's not much clean-up to do. It's easy to avoid waste and clean-up by applying cement carefully and by dedicating

A cut-off brush kept in a jar of brush-grade contact cement is handy for small, quick jobs. Keep enough cement in the jar to cover the bristles, and keep the lid tight. If the cement gets too thick, you can thin it with solvent.

brushes, rollers, and containers to contact-cement use. Brushes can be kept in an airtight glass jar with enough cement to cover the bristles, and rollers can be kept at the ready in a 5-gallon pail with about a gallon of contact cement, a rolling screen, and a tight-fitting lid. For solvent-based cement, use a clean metal pail, and for water-based cement, use a clean chemical-resistant plastic bucket, such as the buckets used for pool chemicals. These set-ups work well for storage even if you're not using contact cement regularly.

To clean up solvent-based cement, try a non-toxic citrus-based cleaner first, then proceed to solvents such as mineral spirits or toluene if you have to. Water-based cement is a breeze to clean up with soapy water when it's still wet, even from roller covers and spray guns. If water-based contact cement dries out, it has to be cleaned up with citrus cleaner or solvents just like solvent-based cement. Flush out spray guns used for water-based cement with a mild ammonia solution after the soapy water, then rinse out the ammonia. If you need

Handy contact-cement application rigs include a roller bucket with a screen and tight-fitting lid, and a cheap spray gun. A plastic bucket is fine for water-based cement; for solvent-based cement, use a clean metal pail.

to discard excess solvent-based or water-based cement, allow it to dry out completely and harden to a non-sticky solid first.

HEALTH AND SAFETY

Although water-based contact cement is much safer to use than solvent-based cement, you should avoid breathing its fumes and avoid skin contact. When using non-flammable solvent-based cement, wear gloves and a respirator and work in a well-ventilated area if possible. The chlorinated solvents used in non-flammable cement can affect your cardiovascular and central nervous systems. Make sure your respirator is rated for chlorinated solvents if your cement contains them.

Flammable cements are toxic and present a great fire hazard, because they contain a high percentage of volatile solvents. When contact cement is spread over a large surface area (which it often is), these solvents can cause explosive fumes to build up to dangerous levels in

the shop as they evaporate from the cement. Wear gloves and a respirator, and exhaust the fumes with an explosion-proof fan.

Performance

Despite its potential hazards and poor cohesion, contact cement is popular because it's convenient to use. In good conditions it dries fast and allows instant assembly without the need for clamping. As with any other adhesive, though, you should be as familiar as possible with contact cement's performance properties so you can use it to maximum advantage.

APPLICATION

Because contact cement has a high degree of tack, it can be hard to brush or roll. Thin it down if you have to, and don't overwork it as you apply it. Keep the applicator wet, brush or roll the cement slowly and steadily, and don't go back over areas you've just coated. Water-based cement is more forgiving to apply than solvent-based cement, because it dries more slowly. Water-based cement is also easier to spray than solvent-based cement. It goes on smoothly in a controlled, well-dispersed pattern, whereas solvent-based cement is flung onto the work as a stringy web by the spray gun. Consult technical data sheets from glue suppliers and spray-gun manufacturers to find the best fluid needle and nozzle sizes for spraying cement.

Contact cement should dry to an even, continuous film on top of the entire gluing surface, free of bubbles, globs of cement, and foreign particles.

If you apply too little cement or apply cement that has been thinned out too much, it may get absorbed by porous gluing surfaces and show bare spots, and you'll have to recoat. If the cement coating is too thick, the bond layer will not be as strong as it should be. On porous surfaces, I get the best results from applying two thin coats of cement rather than one heavy coat. I make sure to wait until the first coat is completely dry before recoating. This is especially important with water-based cement, because the second coat can trap residual moisture from the first coat. When the cement is dry, you should be able to skim your hand lightly over its surface without sticking to it. While you're skimming, knock down any bubbles and sweep off any debris that's settled on the film.

For best results with any cement, use a good roller cover and push the roller steadily at a moderate rate to avoid bubbles. Don't overwork the glue.

DRYING TIME

The drying time of contact cement is governed by three factors: shop temperature, relative humidity, and airflow. The shop should be at least 65°F when you apply cement, and the humidity should be lower than 65% if possible. If you use contact cement in overly cold or humid conditions, it will take longer to dry, and the contact bond will have lower ultimate strength. Also keep in mind that water-based cement takes two to three times as long to dry as solvent-based cement.

Controlled airflow can help reduce drying times. The air should be warm and clean. Don't train the air source on the work until after the cement has been applied, and make sure you take every safety precaution when using any heat or electrical air sources near drying flammable cement.

You don't have to assemble parts the minute that the cement is dry. Most cements have an open time of two or three hours. If you wait longer than that, you'll have to recoat the surfaces and let them dry again.

ASSEMBLY

Once the cement is dry on both surfaces, you can close the assembly. With most cements, you only get one chance to get the parts aligned properly because the cement bonds instantly. You can put dowels or Venetian-blind slats between the two glue surfaces to keep them apart while you align them, then withdraw the dowels or slats one at a time while you press the surfaces together.

PRESSURE

When a contact bond has been assembled, pressure should be applied to the work as soon as possible. Applied pressure is critical to the ultimate strength of the bond because it flattens the cement into a thinner, more uniform layer. Many woodworkers apply pressure with a rubber roller or a wooden block.

The more pressure you use with the roller, the stronger the bond will be. A hand roller provides barely enough pressure; a brief period in a veneer press would be better.

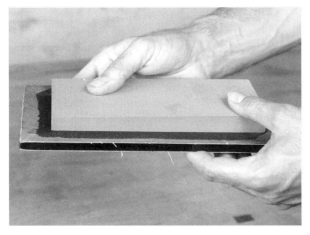

Contact cement bonds sandpaper and glass effectively to make a flattening plate for Japanese waterstones. The cement bond is surprisingly water resistant—it will last indefinitely if the flattening plate is dried off after each use.

This exerts only the bare minimum amount of pressure needed for a good bond. You can dramatically strengthen a contact-cement bond by pressing the work with clamps and cauls or by putting the work in a veneer press. This extra step somewhat diminishes the convenience of using contact cement, but it can be short—even 30 seconds worth of substantial pressure can make a great difference in the quality of a contact-cement glue line.

STRUCTURAL QUALITIES

Much of the strength of a contact-cement bond comes from the materials being bonded. A kitchen countertop overlaid with plastic laminate has a strong, resilient cement bond because the glue surface area is large, and the materials being bonded are rigid. A sheet of bending plywood overlaid with a paper-backed veneer has a weak, flexible cement bond because both materials are pliant. In general, contact cement is not rigid, creep resistant, or strong enough for permanent structural woodworking bonds. Contact cement develops its ultimate strength over a period of days or weeks, depending on shop conditions.

ENDURANCE QUALITIES

Heat resistance is contact cement's one great endurance property. It can withstand much higher temperatures than PVA, which is why it is the adhesive of choice for post-forming, where plastic laminate is pressed around contoured substrate profiles (such as countertop bullnose edges) with heat.

Cured contact cement has low water resistance. Moisture softens the bond layer and degrades its ability to adhere to gluing surfaces, even though the cement won't dissolve in water. Nevertheless, contact cement has surprising longevity when exposed to brief intermittent doses of moisture.

The solvent resistance of contact cement is also very low. This is one reason it's a poor choice for veneering. Finishes applied over the veneer will penetrate its surface and can weaken or dissolve the cement beneath, which will cause the veneer to lift from the substrate.

5

Hot-Melt Adhesives

One of the best lessons I've learned about gluing is that different glue joints have different life expectancies. A treasured piece of custom-made furniture should hold together indefinitely, but a stop block on a jig may only need to be glued down for an hour. There's no point in spending lots of time or using demanding glues to bond assemblies that will have relatively short useful lives.

That's why hot melt has become popular among woodworkers in small shops. For speed and convenience in gluing up temporary assemblies, it's hard to beat, especially since the work can be bonded without clamping. (Occasionally, though, a hot-melt assembly may have to be clamped.)

In a way, however, it's unfair to describe hot melt as strictly a temporary glue. The woodworking industry considers hot melt to be a permanent adhesive because it will produce bonds that hold up satisfactorily over the expected life of various types of furniture and cabinets. Commercial shops use it extensively for gluing veneers and plastic laminates to the faces and edges of cabinet parts cut from flakeboard and other sheet stock. High-volume industrial plants are heavily committed to hot melt as a primary adhesive for certain production operations and typically invest many thousands of dollars in equipment dedicated to hot-melt bonding.

Chemical Basis and Behavior

Hot melt has the most direct and descriptive name in the glue business. It's like a high-tech stick of sealing wax. When you heat it up, it liquefies; when it cools, it solidifies, and it can be remelted

In many small shops, hot melt is most often used in stick form. Hot-melt guns and sticks are available in a quality range from advanced industrial grade to inexpensive hobby grade.

Hot melt is great for temporary bonds. Here, the stop block shown in the photo on p. 56 has been bandsawn off after use. The stop-block remnant and glue layer are easy to remove with heat and a chisel so the jig can be reused for other purposes.

and cooled again if needed. Each of the many different hot-melt formulas available has an optimal melt point and becomes fluid with no significant loss of volume. Once it's applied, hot melt cools and hardens at a rate that's governed by the temperature of the work and the shop.

Hot melt is a very cohesive glue. When it's cold, it's a dense, smooth, uniform solid. When it's heated, it becomes very tacky and is hard to to separate cleanly and quickly, even though it's fluid. Strings of the glue can trail from bead to bead as you apply it, like strands of a spider's web. Hot melt doesn't have outstanding adhesive properties, though. Due to its high viscosity at application and rapid hardening, it doesn't have a chance to penetrate and wet joint surfaces thoroughly. It can also be hard to press into a thin layer, so it can't develop the kind of intimate surface contact that helps create enduring bonds. The thicker the bond layer of hot melt between two joint surfaces, the greater the glue's cohesive power is in relation to its adhesive power, and the more likely the bond layer will be to release from one or both of the surfaces, causing the joint to fail.

Formulation

Hot melt is available in a greater variety of forms than any other glue used in woodworking. With a 100% solids content, it's available in sticks, pellets, slugs, sheets, and pre-applied backing, among others. Hot melt is also the adhesive with the greatest number of dedicated application systems, including hot guns, tacking irons, hot presses, and

low-
...ins,
). 38)
...uch as
...dants,
...d to a
...mula.
...are
...ended
...ds are
...to
...nce

...ryone
...actors,
...des.
...glues
...in all
...las.
...ailable

...ey are
...ar,
...ese
...is
...nding.
...ally

...being
...l hot-
...t was
...n of
...ethane-

...nds. It
...like
...lid after
...reaction.
...fied
...ame
...rties as

polyuretnane glue.

Hot melt in sheet form. After a sheet of hot melt is ironed down and pressed with a block or roller, the paper backing is removed, and veneer (or an overlay) is ironed onto the glue layer.

Buying guns and hot-melt sticks isn't cheap if you use industrial-grade products, and the new reactive hot-melt systems currently cost hundreds of dollars. On the other hand, you can use hot melt in sheet form for the price of a yard-sale clothes iron. Sheet hot melt is manufactured both as an individual product and a pre-applied backing on veneer and plastic laminate edge-banding tapes. Instead of being heated first and then positioned, sheet hot-melt products are positioned first, then heated with the clothes iron or a similar heat source. They allow veneer and overlay work to be done quickly and easily as well as cheaply.

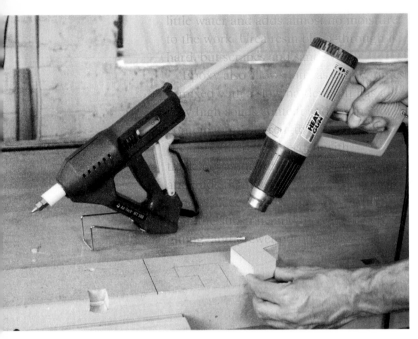

To extend hot melt's open and closed assembly times, use a high-quality hot-melt gun that delivers lots of heat, and pre-warm the work.

SURFACE SENSITIVITY

Because hot melt has fairly low adhesive properties and doesn't wet gluing surfaces very well, it does not bond well to contaminated surfaces. Make sure bonding surfaces are as clean as possible before gluing up so the glue won't release from the work. If you're using hot melt with oily tropical hardwoods, scuff the bonding surfaces with sandpaper instead of wiping them with solvent. Using a solvent can draw oils out of wood pores, leaving residual traces of them on the bonding surfaces.

MELT POINTS

Hot-melt sticks are available in two different melt points. Low-melt glues have melt points in a general range of 240°F to 285°F, and high-melt (or hot-melt) glues are rated at between 350°F and 400°F. The two melt points require separate glue guns or different heat settings on dual-temperature guns. High-melt glues are more commonly available than low-melt glues and are sold in a greater variety. Low-melt glue is used when heat-related safety is an issue (although 265°F would still hurt plenty if you burned yourself). It's also used when the materials being glued can't withstand the higher heat, as is the case with foams and plastics.

For my purposes, high-melt guns and glues are the best, so I use them exclusively to keep hot-melt gluing as simple as possible. Some woodworkers use high-melt guns to apply low-melt as well as high-melt glues. The high-melt gun overheats the low-melt glue, which makes it more fluid and also extends its working time. Even though this doesn't ruin the glue, I don't recommend it. The thin, overheated low-melt glue can backflow out of a high-melt gun's infeed port, burn your hands, and possibly ruin the gun. Backmelt, as it's called, can also be very messy. Furthermore, when you want to switch back to high-melt glue, you first have to purge the gun of the low-melt glue. Dual-temperature glue guns have this same problem.

If you need more fluidity and working time from high-melt glue, you've got two good options. One is to use a higher-quality gun. Good guns produce higher melt temperatures than cheap guns, which is helpful if you use industrial-grade glue, which typically requires a bit more heat than consumer-grade glue. Good guns also produce a lot more heat and recover their heat more quickly as you use them, which means that they will liquefy cold incoming glue as fast as you can squeeze the trigger. Your second option is to use a lower-viscosity glue. Industrial hot melt comes in a variety of melt viscosities, and you should be able to find one that suits your needs.

SIZES

Hot-melt sticks are sold in a variety of diameters; ½ in. and ⅝ in. are the most common. The sizes of the sticks are nominal, so they're not always made to the exact stated dimension. Actual stick diameters also vary from manufacturer to manufacturer, as do the diameters of the infeed ports of glue guns, so there is no guarantee that a ½-in. glue gun will accept every ½-in. stick. If a stick is oversized for a gun's infeed port diameter, it won't feed properly. If the stick is undersized, the hot glue may backmelt out of the infeed port. Some companies resolve this confusion by giving their guns' infeed ports and their glue sticks special matching shapes and sizes, so that they have to be used together as an exclusive system.

Handling and Storage

For all practical purposes, hot melt seems to have an unlimited shelf life if it's stored properly. My 16-year-old hobby-grade glue sticks are still going strong. Hot melt should not be stored in temperature extremes; heat may deform it, and cold will make it brittle. Hot-melt-backed veneer that's improperly stored may cup or stiffen, which makes it much harder to bond to the work when it's applied.

CLEAN-UP

Once cooled, hot melt cleans up easily. Squeeze-out can be peeled off with a fingernail if you catch it before it hardens completely; if not, it can be pared away with a sharp hand tool or scraped off. Wave a hair dryer or heat gun over it first if needed to soften it up a bit. Small amounts of hot-melt squeeze-out can also be sanded off effectively. Solvents such as acetone, naphtha, and toluene, which soften hot melt, are not good choices for cleaning squeeze-out from workpieces. They dilute the glue and drive it into porous wood surfaces, where it can resist stains and finishes, causing a blotchy appearance. Solvents used for clean-up can also migrate into joints and soften hot-melt glue lines.

HEALTH AND SAFETY

The main hot-melt safety hazards are the heat from applicators and the molten glue, both of which can burn you. Hot melt also contains waxes that give off fumes and vapors as the glue reaches its melt point. These emissions aren't highly offensive, but they could cause respiratory tract irritation if you work with hot melt for long periods of time. For protection, you may want to wear a respirator, but simply ventilating the work area should be adequate in most cases.

Performance

Even though the success you have using hot melt depends largely on your equipment, skill is still important. You must be able to apply the right amount of glue as fast as possible, or else the hot

Hot melt is not a suitable structural adhesive for work in solid wood. Here, two blocks bonded by hot melt are pulled apart by hand, with the aid of a clamp.

Hot melt bonds better to surfaces that have some tooth, or roughness to them, which is why the glue works well on the crisply sawn edges of plywood and flakeboard.

WORKING TEMPERATURE

The warmer the shop and the work are, the easier it will be to use hot melt. If your shop is cold, you may want to warm the work with a heat source before gluing. Prewarmed joint surfaces help the glue remain hot once it's applied, which increases its open assembly time and enhances its ability to coat and penetrate joint surfaces.

ASSEMBLY TIMES

Hot melt usually has a lightning-fast closed assembly time because its mass is dispersed into a thin layer when parts are squeezed together, which rapidly dissipates its heat. Prewarming the work extends the closed assembly time because the heat of the glue and the work is contained when the assembly is closed. This allows the glue to be pressed into a thinner layer. Wood holds heat for a long time, so prewarmed hot-melt joints take longer to gain strength than cold joints.

Some hot melts have longer assembly times, but those that have the shortest assembly times often provide the strongest bonds, so you may have to determine which properties are the most important for your work.

STRUCTURAL QUALITIES

In general, hot melt doesn't have enough strength to bond cabinet and furniture joints properly, even though some industrial hot melts produce fairly dense

melt will begin to cool and harden before you can get the parts pressed closely together, and the glue line will be too thick.

When edge-banding a panel, you have to develop a feel for how to use the heat source, whether it's an iron or a dedicated edge-banding unit. If you don't know how much heat to apply and for how long, you may overheat and overwork the glue, or you may have to reheat and rework an underheated glue line, which would make the job take a lot longer.

and rigid glue lines. Likewise, it doesn't fill gaps well on a structural basis even though it physically fills gaps very well. That's because the thicker a hot-melt bond line is, the less stress and shock resistance it has. The strength of hot melt is always being improved, though, and the new reactive hot-melt/polyurethane hybrids are proving to be strong enough for both structural bonding and gap filling.

If you're reviewing technical specifications to find the strongest, toughest hot melts, don't bother to compare tensile strengths as you would with other glues, because that is a measure of adhesion, which is not hot melt's forte. Instead, look for the lowest elongation percentage. Elongation is a measure of how much a hot melt will resist being stretched and pulled apart in controlled tests. The lower its elongation percentage, the more cohesive a hot melt is. This is important because cohesion is hot melt's principal source of strength.

Hot-melt-glued veneers must be finished carefully. Fast-drying sprayed finishes usually work well, but slow-drying finishes such as oil contain solvents that can soften the glue layer.

CURED WORKING QUALITIES

As gummy and stringy as hot melt can be to apply, it's fairly sandable after it hardens. Excess glue tends to clump up a bit, but it doesn't load up the sandpaper.

Hot-melt squeeze-out can be hand tooled with ease, but may cake up on machine cutters that run at high speeds and generate lots of heat, which softens the glue. Hot melt can also cause problems with finishes. Trace amounts that are left on the surface may cause blotches if stain is applied, and the solvents in some finishes may soften hot-melt glue lines and cause them to release from the work. In general, fast-drying

finishes such as shellac and lacquer should be used over hot-melted work because their solvents dissipate before they can soften the glue. Slow, penetrating finishes such as oil are the most likely to cause problems.

ENDURANCE QUALITIES

Even though hot melt does have poor resistance to heat, solvents, and mechanical shock, it is generally a stable, enduring substance. Beads of old hot-melt squeeze-out on jigs in my shop show no signs of yellowing, desiccating, cracking, or becoming tacky. Hot melt is also impervious to insects, bacteria, fungi, and other organisms. These properties have helped make it a popular adhesive for museum conservation work.

6

Urea and Resorcinol Resin Glues

As you are deciding which glues are best to use in various situations, it helps to keep in mind that some adhesives were originally developed in response to the shortcomings of other adhesives. This is the case with urea and resorcinol resin glues. When they were first introduced (over 60 years ago), they were seen as a significant advance beyond natural glues because of their superior resistance to moisture, heat, and microorganisms. Urea and resorcinol resin glues quickly became the dominant adhesives in many fields, including aircraft and boat construction. They also made many new products possible, such as exterior- and marine-grade plywood.

Today, urea resin glue (also known as urea formaldehyde glue, or simply UF) and resorcinol are no longer used as extensively as they once were. Newer adhesives, such as epoxy, polyurethane glue, and cross-linking PVA, also handle tasks that require high strength and endurance. Nevertheless, urea and resorcinol resin glues are still widely used and remain great choices for bent laminating, veneering, and other demanding jobs.

Urea resin glue is sold in two forms: a one-part powder (commonly called plastic resin glue) that is mixed with water and a two-part system that is composed of a liquid resin and a powdered catalyst, also known as a hardener (though technically catalysts and hardeners are different substances). Resorcinol is available only as two-part systems that handle and behave much like two-part urea resin glues.

Chemical Basis and Behavior

All urea and resorcinol resin glues operate in basically the same way. Their main constituent is the resin, which is synthesized by combining

Urea and resorcinol resin glues are known by various names. The urea resin glues at left and center are also known as urea formaldehyde glue, or UF glue (the one-part urea glue on the left is often called plastic resin glue). The resorcinol on the right is also known as phenol-resorcinol glue.

ingredients that are reactive when mixed together, such as urea and formaldehyde. These ingredients are combined under acidic conditions, which induce the reaction.

If left alone, the resin ingredients would continue reacting until they produced an inert solid. But as the resin is being manufactured, the reaction is halted at a given point by buffering the resin to a neutral or slightly basic state. This renders the resin stable temporarily, and it is in this state when you buy it.

When you mix up a batch of glue, the catalyst, which is an acid salt such as ammonium chloride, returns the resin to a state of acidity and renews the chemical reaction between the resin's ingredients, which causes the glue to harden. The polymer molecules produced by the reaction are very large and have a great deal of mass, which creates bonds that are very strong and rigid. The molecules also become cross-linked, which gives the bonds a high degree of moisture resistance.

Formulation

Urea and resorcinol resins are both initially manufactured as liquid solutions carried in a mixture of water and alcohol. Liquid urea resins typically contain very little water in relation to alcohol. Resorcinol resins often contain a fairly substantial percentage of water. Some resorcinol resins are a blend of resorcinol and phenolic resins. The phenol component is added for the sake of economy; it's far less expensive than resorcinol.

The powdered hardener used with these resins consists primarily of a catalyst and a filler, such as walnut-shell flour. The filler is added to provide an appropriate color for gluing wood, to help disperse the powdered catalyst, to improve the workability of the glue mix, and to help prevent crazing of the cured glue line.

One-part powdered urea resin glue is a preblended mixture of hardener and urea formaldehyde resin that has been dried and finely ground. Mixing the

Water-based aniline dyes and inks make effective tints for urea resin glues, such as this batch of one-part "plastic resin glue." Always test colorants with small amounts of mixed glue to make sure that they're compatible.

powder with water reconstitutes the dried resin, which is then catalyzed by the blended hardener. Calling plastic resin glue a one-part glue may seem curious—after all, water has to be added to the powder to make the glue. It's called a one-part glue because all of its principal components are combined in the powder, and water is viewed only as a reconstituting solvent.

Urea and resorcinol resin glues contain formaldehyde, which is a known health hazard. Some formulas contain other components that are also known or suspected hazards, such as furfuryl alcohol, which is added to urea resin glue to help it wet and penetrate surfaces and to help prevent crazing, and the phenols in resorcinol, which can be toxic in concentrated amounts.

Some manufacturers have recently reformulated some ureas and resorcinols to make them safer to handle and use. Unfortunately, these updated formulas usually don't perform as well as the older, more hazardous formulas. Some woodworkers insist on using the more hazardous glues because they aren't

willing to give up their superior performance and because the updated formulas aren't completely hazard free themselves. All ureas and resorcinols should be used with proper precautions, as described on pp. 67-68.

COLOR

When mixed, urea resin glues have a tan color, which blends in well with many woods. They can be tinted with either water-based or alcohol-based aniline dyes if needed to help make glue lines as inconspicuous as possible. Resorcinol glues, on the other hand, have a dark purple-brown color that makes even the tightest glue lines highly visible, such as on the edge of a sheet of exterior plywood.

SURFACE SENSITIVITY

One-part urea resin glues may not always bond successfully to dense, oily, or resinous woods. Two-part urea resin and resorcinol glues work better with these woods because they contain alcohol, which helps the glue penetrate and wet the surface of the wood. Heating clamped assemblies also improves glue penetration, as do additives that enhance wetting ability (these are available from some suppliers). Each species is different, so if you have concerns, make test joints to see if adhesion problems occur. Prepare the surfaces of any difficult woods carefully before gluing them up (see pp. 105-107).

MOISTURE CONTENT

Two-part urea resin glue has a lower moisture content than one-part urea resin glue and resorcinols because its liquid resin component is primarily alcohol based. Liquid resorcinol formulas contain some water along with alcohol, which gives them a higher moisture content. One-part powdered urea resin glues have the highest moisture content of the three because they're mixed with water, and consequently they can transfer a lot of moisture to glued work. This added moisture can take several days to evacuate fully from the pieces being glued and might cause unwanted dimensional change in the work.

PERCENTAGE OF SOLIDS

Urea and resorcinol resin glues have a solids content that ranges from 50% to just over 70%. One-part urea resin glues (once they are mixed with water) are at the lower end of the range, resorcinols are in the middle, and two-part urea resin glues are at the high end. Compared to other types of glue, these glues fall above PVAs and below epoxy and polyurethane glue on the solids-content scale.

Handling and Storage

In small shops like mine, urea and resorcinol resin glues have a tough role to play. They are typically used for critical project components that depend on top adhesive performance, such as veneered panels. But because these glues are generally not used as everyday, all-purpose adhesives, they may sit on the shelf for extended periods of time between glue-ups. They have to be stored and handled properly so they will deliver optimal results whenever they're called upon.

Veneering just one face of this ½-in. MDF panel using one-part urea resin glue will introduce to the panel roughly the amount of water shown. Plan your work carefully to avoid moisture-related problems.

SHELF LIFE

Because urea and resorcinol resin glues contain a temporarily suppressed chemical reaction, they have a built-in reactive potential and a fairly short shelf life. To maximize shelf life, store them as carefully as you can. The main concern with liquid resins is heat. The hotter the temperature they are stored at, the shorter their shelf life (see the chart on p. 64).

As liquid resins age, their components begin to react with each other in the container even though no catalyst is present. The resins gradually thicken and eventually solidify. If you have some resin that's thickening, you may want to use it up while you can. Batch mixes made with thickened resin will likely have a shorter pot life and allow less working time than usual.

One-part powdered urea resin glue must be protected from both heat and humidity. If the powder is exposed to humidity, it will react inside the storage container and become unusable. It can be hard to tell if this has happened, though, because spoiled powder often looks the same as fresh powder. The best

Shelf Life

Storage Temperature	One-Part Urea Resin (Powder)	Two-Part Urea Resin (Liquid Resin Only)	Resorcinol (Liquid Resin Only)
stored at 70°F	8 to 12+ months	6 to 9 months	10 to 12+ months
stored at 90°F	6 to 8 months	3 months	4 to 10 months

The shelf-life listings are approximate; actual shelf life varies, depending on conditions and the brand of glue being stored.

Do anything you can to keep moisture away from powdered urea resin glue, especially if you don't use up your supply quickly. Bags and twist ties are simple and effective.

way to check the condition of older one-part urea glue is to mix up a small test batch. If it doesn't mix or cure properly, don't use it.

The powdered catalysts sold with two-part urea and resorcinol systems are very stable. If stored under good conditions, they will keep indefinitely.

PREPARATION

Mixing urea and resorcinol resin glues requires the careful measurement of parts: powdered glue and water with one-part glues, and powdered hardener and liquid resin with two-part glues. You can measure by weight or by volume. The proportions are different, depending on which method you use. Measuring by weight is more accurate, and measuring by volume is more convenient.

Once you've mixed one-part powdered urea resin glue with water according to the recommended proportions, you'll know what its mixed consistency should be like. You can then mix it by eyeballing the proportions if you want. With a two-part urea resin or resorcinol glue you should always measure out the parts before mixing.

Mixing one-part urea resin glue

As you mix one-part urea resin glue, you can't determine the ratio of catalyst to resin, because those components are premeasured and blended into the glue powder. What you can determine is the viscosity of the glue mix, based on the amount of water you add to the glue powder. When mixing the glue, use only high-quality water (see pp. 103-104). Warming the water slightly is helpful. Don't use extremely hot water, though; it will either make the glue mix react too quickly or ruin the mix entirely.

There are two ways to mix one-part urea resin glues. You can add powder to water or water to powder. If you add powder to water, the water must be stirred continuously while the powder is carefully added, or else the glue will become lumpy. Adding water to powder is much easier. Start by adding a small amount to the powder and blend it into a thick, smooth paste. Then thin the paste by adding the rest of the water while stirring the mixture. If you're mixing by eye, add water sparingly to the paste. You can easily end up with glue that's too watery, in which case you'll have to thicken it. If you do have to thicken up some glue, make a separate second thick paste and add this to your watery batch.

When you mix one-part urea resin glue, you can add an extender such as wheat flour if you wish. This is done for the sake of economy because it allows a batch of glue to cover more surface area. I don't recommend adding extender, though. It makes mixing more complex and it lowers the percentage of adhesive solids in a batch mix, which reduces the cured strength and durability of the glue.

Mixing two-part resin glues

As you mix two-part urea or resorcinol glues, you can't control the viscosity of the mix the way you can with a one-part urea glue, but you can control the ratio of resin to hardener, which is more useful. Altering the mix ratio changes both the reaction rate of the glue and its ultimate properties. If you use less hardener, the glue will react more slowly and produce a more resilient bond. If you use more hardener, the glue will react more quickly (which shortens the pot life, assembly times, and curing period) and will produce a more brittle bond. Consult with a supplier or the

Mixing one-part urea resin glue by volume is simple with measuring cups. After you have a feel for the proper recommended consistency of the mixed glue, you can eyeball your proportions if you wish.

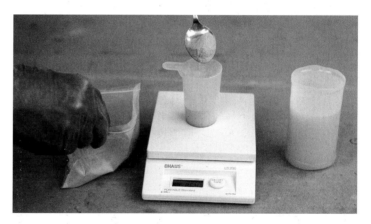

You can mix two-part urea resin glue by weight (preweighed resin is on the right). The digital scale makes measuring small amounts and varying mix ratios a snap. For larger batches, use a postal scale.

manufacturer before mixing glue if you have questions about these ratios.

When mixing two-part urea or resorcinol systems, you can add the powdered hardener to the liquid resin or add the liquid to the powder. The powder is readily assimilated by the resin, so lumping isn't a problem if you mix carefully.

Pot Life

Temperature	One-Part Urea Resin	Two-Part Urea Resin	Resorcinol
65°F	4½ hr.	2 hr.	4¼ hr.
75°F	3¼ hr.	1 hr.	3 hr.
85°F	1¾ hr.	¾ hr.	1¾ hr.
90°F	1 hr.	½ hr.	1 hr.

The times listed above are approximate, and are based on average mix ratios.

POT LIFE

The pot life of mixed urea and resorcinol resin glues is determined by temperature and mix ratio. Two-part urea resin glues are the most reactive (see the chart above). Once a two-part glue is mixed, its catalytic reaction can become exothermic, or heat-producing. This occurs when a large volume of glue is held in a concentrated mass, such as in a deep, narrow mix pot. A plastic 2-lb. yogurt tub full of mixed glue can get hot enough to melt the container and burn you, and it gives off noxious fumes. The heat will also speed up the reaction rate of the mix, and it will polymerize before you can apply it.

To avoid this problem, disperse the volume of glue by transferring it into a wide, shallow container soon after mixing. Another good strategy is to mix glue in several smaller batches rather than one large batch, if possible. You can also extend the pot life of a batch of glue by refrigerating it right after mixing.

CLEAN-UP

Once you've applied urea or resorcinol resin glue, clean-up is easy. Hands and tools can be cleaned with soap and water. Don't use scalding hot water—its heat can gum up or solidify the traces of excess glue you're trying to clean up.

There is no single best method for cleaning excess glue from the work itself. It depends on how heavily the assembly will be worked after the glue dries. When gluing joints that are surrounded by previously prepared show surfaces, such as a table leg-and-apron joint, I clean the excess with water, rags, and brushes while the glue is still wet. With bent laminations or veneered panels, I let the glue cure and then remove the excess, using power tools and machinery as much as possible.

Resorcinol squeeze-out can cause problems on some woods because of its dark color. If you are gluing up wood that's light colored and porous, like white oak, cleaning up the squeeze-out with water may drive glue residue down into the pores and darken them or may stain the surface surrounding the glue joint.

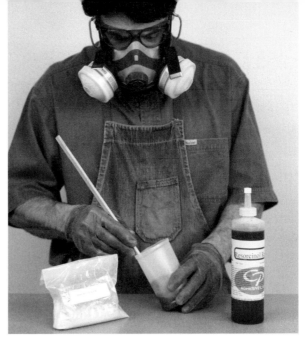

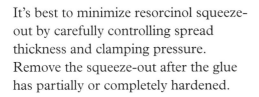

There's no need to pour out unused glue from a batch mix. If you use plastic mix pots, just let the glue cure to a solid and it will release from the pot for easy disposal.

Safety gear is important when mixing and using urea and resorcinol resin glues. Glasses protect against splashing, the respirator guards against airborne powder and fumes, and rubber gloves prevent contact with the skin.

It's best to minimize resorcinol squeeze-out by carefully controlling spread thickness and clamping pressure. Remove the squeeze-out after the glue has partially or completely hardened.

DISPOSAL

Avoid washing glue down the drain wherever possible. It's best to dispose of glue as a cured solid instead of separate unmixed ingredients. Unused glue that's exceeded its shelf life should be mixed and cured before being discarded. Even if the glue doesn't cure perfectly, it's still a more benign waste product than unmixed glue. Cured glue will release neatly as a block from the inside of plastic containers, and you'll be able to use them as mix pots again.

HEALTH AND SAFETY

Ureas and resorcinols should be used with the utmost attention to health and safety concerns. Inhalation and contact are both considered hazardous, because these glues contain potential carcinogens, and repeated exposure can cause sensitization or serious skin and respiratory illnesses. Eye contact causes irritation and even loss of vision.

Use caution when mixing and spreading urea or resorcinol resin glue and when processing glued work after it has cured. Mixing the glue involves exposure to airborne particles as well as hazardous fumes, so a simple dust mask won't do the trick. Use an organic vapor respirator, and wear rubber gloves and eye protection.

When you apply a urea or resorcinol resin glue, your risk is determined by the size of the surfaces you're gluing: The bigger the surfaces, the greater the exposure. If you're roll-coating the faces of large panels for laminating or veneering, you may want to wear protective clothing and an air-supplied full-face breathing helmet instead of just a respirator and glasses.

HAZARDS OF CURED GLUE

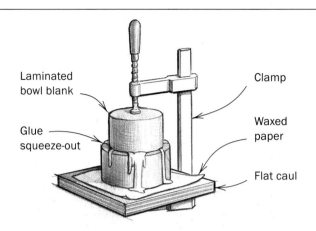

Urea and resorcinol resin glues cure to a hard film that can be dangerous. Here, the squeeze-out from a stack-laminated bowl blank flows down onto a sheet of waxed paper and dries as a small puddle with sharp edges.

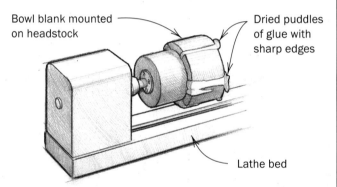

When the bowl blank is mounted on the lathe and spinning, these dried puddles of squeeze-out become highly dangerous cutters if not removed prior to mounting. They can easily slice fingers and wrists to the bone.

Once urea or resorcinol resin glue has cured, it will off-gas formaldehyde fumes. Most of the fumes dissipate within the first day after curing and gradually diminish to negligible amounts after that. Sheet goods such as flakeboard that contain lots of formaldehyde-based adhesive resins can off-gas fumes for months or years. When you machine through cured urea or resorcinol resin glue, whether in work you've glued or in manufactured sheet goods, you should protect yourself against vapors and airborne particles by wearing a respirator. You can reduce your exposure to these hazards by using formaldehyde-free sheet goods, which have become available in the last several years.

Cured excess glue itself can be physically hazardous. Remove it from the work as soon as you can. The squeeze-out can be brittle, and may fracture and fly about as you cut through it (see the drawing at left), so wear safety glasses—even when hand scraping the squeeze-out from an edge joint.

Performance

Once mixed, most ureas and resorcinols are easy to apply. They flow readily and spread evenly without overworking because they don't have much initial tack. This makes them good choices for a variety of work. They flow around the surfaces of enclosed structural joints well, and they spread well on flat surfaces for face laminating and veneering. Glue-coated parts are virtually tack free, so they are easy to assemble and position for clamping.

Because these glues spread well, you can easily apply too little or too much glue, so it's important to control the application. If you apply too little glue, you may end up with a starved joint. If you apply too much glue, you'll get excessive squeeze-out and will have to spend more time cleaning up. Applying too much one-part urea resin glue may also add excessive amounts of moisture to your work.

Tips for Success with Urea and Resorcinol Resin Glues

Even though urea and resorcinol resin glues aren't hard to use, they're also not very forgiving adhesives. To work well, they have to penetrate joint surfaces, but in a controlled manner. When they don't penetrate properly, they don't adhere well. On the other hand, they can be overly absorbed by glued surfaces, which starves the bond layer. Either way, the cured joint will fail.

Here are some strategies for avoiding such problems:

• Check the moisture content of the wood being glued to make sure it's within the proper range. One-part urea requires wood with a moisture content of 6% to 12%, two-part urea requires a moisture content of 7% to 15%, and resorcinol requires a moisture content of 6% to 16% (see the chart below).The ideal moisture content falls in the middle of these ranges. If the moisture content is too low, the wood will absorb too much moisture from the glue. If it's too high, the glue will overpenetrate the wood and become diluted. Glue that's either dried out or watered down won't bond properly.

• Consider the species being glued. Soft, porous, wood may absorb lots of moisture from the glue, regardless of how dry or wet that wood is. Try prewetting porous surfaces with glue to offset the absorption and then recoat them with glue before assembly.

• Coat both parts of a joint with glue to ensure that all surfaces will be moistened before assembly. Don't do this when gluing veneer, though. Put glue only on the substrate, and leave the veneer uncoated prior to pressing it on the substrate.

• Regulate the spread thickness for assemblies with large surface areas. The larger the surface, the more critical it is to control the amount of glue you spread. A veneered tabletop needs a more carefully controlled spread than a lap joint.

• Don't rush assemblies unless necessary, especially with dense species. If parts are glue-coated, closed and clamped in rapid succession, the glue may not have a good chance to penetrate and wet joint surfaces before being driven out by clamping pressure.

• Don't overclamp assemblies. These glues are easy to squeeze from clamped glue joints. Excessive clamping pressure will leave too little moisture and/or glue in the joint.

Wood Moisture Content			
	One-Part Urea Resin	**Two-Part Urea Resin**	**Resorcinol**
Minimum	6%	7%	6%
Ideal	8% to 10%	9% to 10%	8% to 12%
Maximum	12%	15%	16%

Open Assembly Time

Shop Temperature	One-Part Urea Resin	Two-Part Urea Resin	Resorcinol
70°F	40 min.	30 min.	15 min. to 25 min.
80°F	30 min.	20 min.	10 min. to 15 min.
90°F	20 min.	10 min.	7 min. to 10 min.

The times listed above are approximate and are based on average mix ratios. If you alter the mix ratios for two-part glues or if your shop is colder than 70°F, you'll have even more time to spread glue and assemble work.

Factors Affecting Closed Assembly Times of Urea and Resorcinol Resin Glues

Longer closed times	Shorter closed times
Thick glue spread	Thin glue spread
Humidity at 50% or above	Humidity below 40%
Shop temperature below 70°F	Shop temperature above 70°F
Glued parts assembled quickly	Glued parts left open momentarily

One reason woodworkers use urea and resorcinol resin glues is because they have fairly long open and closed assembly times (see the charts above). Closed assembly times are variable and depend on spread thickness, shop humidity and temperature, and the length of time that glue-spread parts are left unassembled in the open air. A thicker spread, higher humidity, lower temperature, and rapid closing of an assembly after gluing all help extend the closed assembly time.

CLAMPING PRESSURE

Urea and resorcinol resin glues require only moderate clamping pressure for unstressed assemblies. With their low tack, they squeeze out readily, and excessive clamping pressure can drive too much glue out of a joint. For tensioned glue-ups such as bent laminations, use whatever pressure is required to pull the work snugly to the bending form.

Clamp Time

Shop Temperature	One-Part Urea Resin	Two-Part Urea Resin	Resorcinol
70°F	5 hr. to 10 hr.	4 hr.	8 hr. to 10 hr.
80°F	4 hr. to 8 hr.	2½ hr.	4 hr. to 6 hr.
90°F	3 hr. to 5 hr.	1½ hr.	2½ hr. to 3 hr.

The times listed above are approximate, and are based on average mix ratios. If you add more hardener to a two-part glue or heat the work beyond 90°F, clamp times can be reduced further.

CLAMP TIME

Once glued work is clamped, it has to be kept at a temperature of at least 65°F to 70°F to set and cure properly. If your shop is cold and you have to heat the work so the glue will cure, you may want to warm it beyond the minimum, say to 80°F or 90°F, to speed up the glue's reaction rate and reduce the required clamp time (see the chart above).

Warming up a clamped assembly doesn't have to be an elaborate procedure. You can use common heat sources such as those shown in the photo on p. 112. These heat sources must be used with extreme caution, though. Even the simplest heating arrangements can be hazardous if not properly thought out, set up, and monitored constantly.

Unstressed work such as flat veneered panels can be unclamped and left to cure when the glue has set to a continuous rubbery film. Save a little excess glue from the mix, and use that to check the hardness of the curing glue. (Don't use a large volume of glue for this, because it may become exothermic and cure before

When gluing with resorcinol, don't overtighten the clamps, or you may drive too much glue from the joint. Aim for a controlled, minimum squeeze-out to keep the glue from staining light, porous wood such as this white oak.

Monitor the hardness of curing urea or resorcinol resin glue by testing a small amount of excess in a shallow dish (less than ⅛ in. deep). When the sample hardens to a rubbery state, the clamps can be removed from the glued-up assembly.

the glue in the assembly does.) Stressed assemblies such as bent laminations should be left clamped until the glue has completely hardened.

STRUCTURAL QUALITIES

Urea and resorcinol resin glues cure at an even, consistent rate throughout their mass because they polymerize by chemical reaction. Despite long clamp times, they reach full strength fairly quickly if cured under ideal conditions. They cure to hard, strong, and rigid bonds, which make them highly creep resistant. These bonds can become crazed or brittle, though, which is why high-quality glues have craze-resistant additives.

GAP-FILLING ABILITY

A glue's ability to fill gaps well is based on how much it shrinks while it cures. One-part urea resin glue shrinks significantly because it's mixed with water and gives up a lot of moisture as it cures. Two-part urea and resorcinol glues cure with virtually no shrinkage and have a high solids content, so they are good for filling small gaps. These glues cure to such hard bonds that they can fracture if used to fill large gaps, however. If you have to fill gaps with a glue, the best strategy is to modify a two-part urea resin glue with PVA (see the sidebar on the facing page).

CURED WORKING PROPERTIES

Steel tool edges chip or dull rapidly when cutting through cured urea or resorcinol resin glue. Use carbide blades and cutters as much as possible when machining glued work. The cured glue is very sandable, though, so power abrasives are a good choice for removing squeeze-out and surfacing glued work. Make sure all excess glue is removed from show surfaces before applying finishes because any glue left on the work will resist stains and coatings.

ENDURANCE QUALITIES

Endurance is where the difference between ureas and resorcinols is most evident. There's no disputing resorcinol's superior resistance to moisture, heat, and shock. Many glues claim to be waterproof these days— resorcinol is one of the few that has a legitimate right to make that claim. That's why it's used extensively in exterior- and marine-grade plywood and for challenging jobs where complete water resistance is crucial.

Modifying Urea Resin Glue with PVA

If you want to compensate for some of the shortcomings of urea resin glue, you can modify it with PVA glue, which can be added in ratios all the way up to 1:1. Adding PVA glue to urea resin glue plasticizes the cured bond, which makes it more durable. It also shortens the clamp time and improves the ability of urea glue to "bite in" (penetrate) when gluing hard, dense woods such as birch or maple. Not all PVAs are compatible with urea glues, so mix test batches to check compatibility. You'll be able to tell if two glues are incompatible by the appearance of the mix. Also, test various mix ratios to find those that suit your needs. Finally, when using PVA glue as a modifier, remember that it has a higher moisture content and lower moisture resistance than either one-part or two-part urea resin glue. Those differences may make PVA an unsuitable modifier in some cases.

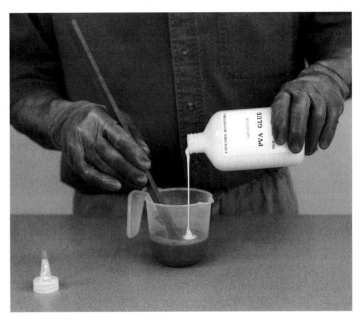

You can modify urea resin glue with PVA. Once you make sure that the PVA you're using is compatible with the urea resin glue, add it in small amounts at first and compare cured batches to find the best mix ratios for your work.

Urea resin glue resists occasional exposure to moisture (including full immersion in water) very well. The key word is "occasional;" urea resin glue will break down if exposed to continual moisture. Urea resin glue also has only moderate heat resistance. A combination of heat (including direct sunlight) and moisture will break it down readily, so it's not the best glue for boat cabin hatches or trivets. It also has limited shock resistance, unless modified with PVA as discussed in the sidebar above.

The original head of this mallet was laminated from dense, oily tropical wood with one-part urea resin glue, which has limited shock resistance and penetrating ability. The new head was glued up with a more durable industrial-grade PVA.

7

Epoxy

Epoxy has been revolutionizing the way that people design and build things for over 50 years, and is now used regularly in most woodworking shops. Today, epoxy is available in many different forms and is used as a sealant, consolidant, coating, and casting material as well as an adhesive.

Chemical Basis and Behavior

Epoxy cures by means of a chemical reaction between a resin and a hardener, which are both stable individual compounds. When they are combined, they are complementary partners in the ensuing chemical reaction and integral elements of the polymerized solid.

Epoxy's resin and hardener components both have essential roles in determining its performance and ultimate properties. They also react with each other in a very thorough, efficient manner. Epoxy mixes lose only about 1% of their volume as they polymerize and are chemically and dimensionally stable once they have completely cured.

Formulation

Most of the epoxies that woodworkers use are produced by small companies that buy basic components (such as resin and hardener) from large chemical manufacturers. These small companies formulate their epoxy from a choice of a dozen or so different resins, which are matched up with a choice of hundreds of different hardeners, modifiers, and additives to suit specific technical requirements. Thousands of finished formulas are possible based on different combinations of these components.

COLOR

Most epoxies are clear or amber, and produce glue lines whose appearance may not blend well with the materials you're gluing. If that's the case, you can tint epoxy as described on pp. 79-80. Most epoxies also yellow with age, so if you need long-term color stability for bonding materials such as glass, use a non-yellowing, high-clarity epoxy.

GRADE

Epoxies are formulated in consumer-grade and industrial-grade versions. Epoxies that you get at the hardware store are mostly consumer grade and have lower ultimate properties, such as strength and endurance, than the epoxies that are sold by aircraft, marine and woodworking specialty suppliers. The consumer formulas are still strong enough for many woodworking applications, though.

VISCOSITY

Because epoxy is used for many different functions, it is sold in consistencies ranging from water-thin liquids to heavy pastes and putties. Many of these are not well suited for use as woodworking adhesives unless you modify them. For example, some marine epoxies are too thin to be used as woodworking adhesives and must be thickened with additives before being used as a glue. Thickening an epoxy is not a big chore, but if you don't want to bother with additives, you can buy higher-viscosity epoxies and gels instead.

HARDENERS

Some epoxies are available with a choice of two or three different hardeners, which react with the resin at different rates. Having a choice of hardeners

These church doors, built by the author, have torsion-box cores with laminated solid wood faces. Epoxy was used exclusively throughout as an adhesive and as a sealer under the varnish finish. (Photo by Lance Patterson.)

makes epoxy much easier to use when you have gluing jobs of differing size and complexity and work in varying temperature and humidity conditions, which can affect epoxy's performance.

HARDENING SPEED

Epoxy formulas are available in a wide range of hardening speeds. I bond glass with an epoxy that takes two days to harden. But there are also fast-curing "five-minute" epoxies, some of which are too fast for any use other than spot bonding. Most five-minute formulas

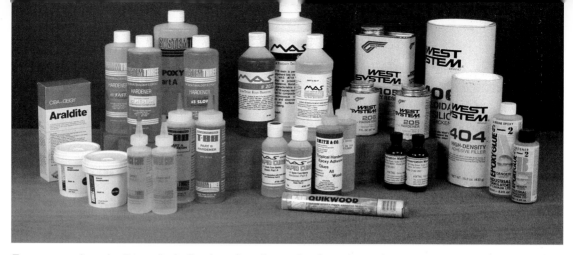

Epoxy comes in many forms, including low-viscosity marine formulas, gels, pastes, and the "cookie dough" roll in front. The marine epoxy systems in the rear offer a choice of hardeners and off-the-shelf additives.

Epoxy excels at bonding difficult-to-glue woods such as dense, oily, resinous tropical hardwoods, as well as many other materials, such as metal and glass.

have lower ultimate properties than epoxies that cure at regular speeds. Some new fast-curing formulas, however, are nearly as strong and durable as regular-speed epoxies.

SURFACE SENSITIVITY

Part of epoxy's great versatility is its ability to bond many different types of materials. Some formulas are designed specifically for bonding certain materials, such as metal, stone, glass, and tropical hardwoods. However, epoxy will not bond properly to some of these materials unless their surfaces have been carefully prepared beforehand. Different materials require different surface preparation routines. For example, acidic materials such as bone should be neutralized with an alkaline solution. Metals such as aluminum, brass, and steel should be rinsed with various chemicals to remove oxide layers. For more information on materials and surface preparation of non-wood materials, see pp. 158-159.

If this seems like more effort than you care to invest in surface preparation, at least make sure that your bonding surfaces are mechanically clean before applying epoxy. If needed, wipe surfaces with a solvent to remove contaminants such as dirt and oil. Then roughen, or tooth, any smooth surfaces with a tool or sandpaper. Epoxy doesn't bond well to surfaces that are dead smooth, such as hand-planed wood or polished metal. With some materials (like many species of wood), scuffing or toothing is the only surface preparation that's ever really needed. With other materials, solvent wiping and scuffing are done along with (and prior to) the chemical rinses.

MOISTURE CONTENT

Epoxy resins are 100% solids, and any moisture in an epoxy mix is carried by the hardener. Any moisture that epoxy attracts after it's mixed is due to excess hardener in the mix. Epoxy's minimal moisture content allows it to cure with negligible shrinkage, which is one reason it's such a good gap filler.

Handling and Storage

Epoxy is more convenient to work with than many people realize. It tolerates less-than-ideal storage conditions better than one-part glues such as polyurethane and cyanoacrylate, and it's easier to mix than one-part urea resin glue or casein glue. Once mixed, epoxy is highly versatile, but it has to be handled carefully because it's a toxic substance.

SHELF LIFE

Epoxy has a good shelf life if protected from temperature, light, and moisture extremes. Ideally, it should be stored between 50°F and 90°F. Stored at temperatures below 50°F, epoxy will thicken considerably and become hazy

and crystallized; it will also freeze. If it freezes or crystallizes, you can restore it by placing its container in a hot (120°F) water bath, just as you might do for a jar of crystallized honey. Epoxy components shouldn't be routinely subjected to rapid temperature cycling, though, because condensation could form inside partially full containers. Stored resins will generally resist added moisture such as condensation fairly well, but hardeners can be ruined by exposure to moisture.

Properly stored epoxy can last for years without deteriorating, though many hardeners will yellow with age. Considering the cost of epoxy, if it works, I'll use it, no matter how old it is. However, I make small test batches before using older epoxy to make sure it still mixes and cures properly.

PREPARATION

Epoxy requires careful measuring of parts before mixing, but in practice, measuring is simple. Most epoxy is

Cracks in wood can be filled with five-minute epoxy. Once the cracks are filled, this blank will be turned into a tool handle.

Hand pumps are convenient and easy to use, especially when you're working with epoxies that have an unequal mix ratio.

measured by volume instead of weight. In many epoxy systems, the resin and hardener have different tints, which can make it easier to gauge volume proportions in a mix pot. If you use clearly marked graduated containers, you can measure and mix batches of epoxy by eye with speed and accuracy.

That's not to say that mistakes can't be made, though. A common mistake is to reverse the proportions of resin and hardener when mixing epoxies that have an unequal mix ratio, such as 2:1. Another mistake is to measure accurately, but in the wrong ratio, say 3:1 instead of 2:1. This can happen if you use several epoxies that each have different mix ratios.

Some epoxies are sold in containers that make measuring parts almost foolproof because the containers automatically dispense resin and hardener in correct proportions. Consumer-grade epoxies, for example, are widely available in convenient twin-plunger syringes. Similarly, some marine-grade epoxies are sold with

regulated hand pumps that allow you to dispense a precise, consistent amount of material with each stroke. These pumps are especially helpful if you're using epoxy that has a highly disproportionate mix ratio (such as 4:1), or that is packaged in unwieldy containers, such as metal cans.

As convenient as these dispensing systems can be, they also have drawbacks. The tips of the twin plunger syringes are hard to keep clean and they can cake up with hardened epoxy. Epoxy also tends to have a shorter shelf life when packaged in the syringes. The disadvantages of hand pumps are that they can clog, they may gum up when seldom used, and they won't pump material that has thickened in cold storage conditions. Pumps also may not dispense the proper amount of material per stroke. You can check pumps for accuracy by dispensing a full stroke each of resin and hardener into separate graduated containers to verify the respective amounts of material dispensed. Check them when you first install them, then occasionally thereafter.

If you don't dispense resin and hardener accurately, you can't control the mix ratio, which is important. When you mix an epoxy in specified proportions, you can depend on the manufacturer's technical data as an accurate guide to all of its properties, from pot life to cured strength.

Some epoxies do allow slight variations in mix ratio. Epoxies with 1:1 and 2:1 mix ratios will tolerate up to about 10% extra hardener or 20% extra resin in a mix (five-minute epoxies allow only 10% of either). Epoxies with mix ratios of 3:1 and above will tolerate only about 5% extra hardener or 10% extra resin. If you mix with extra resin, the epoxy will have a higher viscosity, a longer working time, and a harder cured bond. If you mix with extra hardener, the glue will have a lower viscosity, a shorter working time, and a softer cured bond.

Proper mixing of epoxy is just as important as measuring. The resin and hardener must be completely blended to ensure that the mix will cure properly. Stir the ingredients thoroughly in the mix pot, making sure that all the material on the bottom and sides is incorporated. Stir briskly to give yourself as much working time as possible after mixing.

ALTERATION

No adhesive can be altered in as many different ways as epoxy. You can alter its viscosity, color, or mix ratio, for example, or you can use it as a vehicle for supplementary ingredients.

When altering viscosity, don't try to thin down a thick epoxy. That's best done with special diluents by epoxy formulators. However, thin epoxies are routinely thickened for various uses. Fillers such as glass fiber and microballoons are added when epoxy will be used as a putty or fairing compound. Fillers such as wood flour and thickeners such as colloidal silica are good additives to use when thickening epoxy for use as a glue.

Since epoxies are clear or light amber and don't blend readily with wood hues, you may want to tint them for use with wood. Tinting may also be necessary for epoxies that have been thickened with silica, which turns them a foamy, opaque white that is even more conspicuous in glue lines. You can tint epoxy by using wood flour as a thickener, or you can use dry pigments or alcohol or water-based aniline dye powders as colorants. The

Always mix epoxy and hardener thoroughly. With fast-acting formulas, mix swiftly as well, especially if you're adding tint and thickener, as shown here. Safety gear is always in fashion for epoxy users.

dye powders don't dissolve in the epoxy—they become suspended in the mix like pigments. Add as little dry colorant as possible to avoid plasticizing the epoxy mix. Universal tinting colors (UTCs) are also an effective way to add color.

My favorite colorant is a predispersed tint that's sold in tubes (see the photo above). It's highly concentrated and mixes readily into epoxy, so adding a quick dab to a batch mix is very convenient and doesn't use up minutes of precious pot life. The tint is dispersed in a plasticizer, however, so if you add too much of it, the epoxy will cure to a solid with lower ultimate properties.

Many other materials besides fillers, thickeners, and colorants may be combined with epoxy. Such additives might include carbon fiber for structural

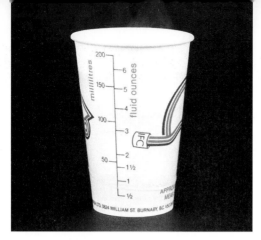

Some mixed epoxies can become very exothermic. This smoking batch mix is so hot that it would burn you if you handled it directly. Use heat-resistant gloves, pliers, or tongs to handle hot batches.

strength or buckshot for weight and stability. Make sure any additives you use are clean, non-moisture-bearing, inert, and non-acidic. Mix small test batches to make sure that epoxy altered with such additives performs and cures properly.

POT LIFE

The pot life of epoxy is one of its most important properties. Epoxy's chemical reaction is influenced by temperature, and it cures faster in hotter conditions. Because many epoxies are also exothermic, a batch mix can produce enough heat during polymerization to flash-cure itself prematurely without warning. If the heat created by the chemical reaction doesn't solidify a batch, it will at least accelerate the curing reaction, which will reduce the amount of available assembly time. Epoxy produces significant exothermic heat only when the glue is in a concentrated mass, such as in a deep, narrow mix pot. If the epoxy mass is dispersed, either onto the work or into a broad, shallow container, heat doesn't build up, the reaction rate slows considerably, and the epoxy maintains its usual working time.

What makes pot life tricky to manage is that it can fluctuate because it's influenced by lots of variable factors (see the chart on the facing page). Many of these factors are interrelated, and some of them, such as humidity, are hard to control, so the best way to regulate pot life is to balance factors that prolong pot life with factors that decrease it. For example, if your shop temperature is over 80°F and you're going to use lots of additives, mix smaller batches, use a slower hardener, transfer batch mixes to a wide, shallow container, and perhaps refrigerate them with cold packs. If you're mixing small batches of epoxy in a cold shop, use a faster hardener and let the mix warm up a bit in the mix pot before dispersing or applying it.

CLEAN-UP

Even though epoxy can be messy to work with, clean-up is a snap if done properly. Use disposable gloves and cheap applicators, and throw them away after use rather than cleaning them with solvents, such as acetone. Don't bother to clean out plastic mix containers. Instead, let the epoxy cure to a solid mass, then pop it free. Epoxy releases well from most plastic containers and from plastic applicators such as squeegees. If it doesn't release from a container, leave it in place, and mix the next batch right on top of it. Never dispose of epoxy resin or hardener as separate components. Mix them together and let them cure to an inert solid, which is a benign waste product.

To remove squeeze-out from your work, let the epoxy cure completely, then remove the excess with hand tools or

Factors Influencing Pot Life

Longer pot life	Shorter pot life
Slower hardener used in mix	Faster hardener used in mix
Fewer additives included	More additives included
Cold, damp shop	Hot, dry shop
Cold resin and hardener	Warm resin and hardener
Small batch volume	Large batch volume
Shallow, wide mix pot	Deep, narrow mix pot
Refrigerated batch mix	Room-temperature batch mix

power equipment. You can also use solvent-based epoxy removers to clean up cured and partially cured epoxy. If you clean wet excess epoxy off the work with solvents, you'll dilute the epoxy and drive it into the surface of the wood. It's very hard to remove all traces of wet epoxy from a bare wood surface, and this can cause problems with staining and finishing. I try to avoid using epoxy removers as much as possible. They don't work on all epoxies, and they contain hazardous chemicals such as methylene chloride.

HEALTH AND SAFETY

Gloves, safety eyewear, and a respirator are important when working with epoxy, because repeated skin contact or inhalation of fumes can lead to sensitization (which is a permanent condition). Once you become sensitized, you'll experience a skin reaction whenever you work with epoxy. These reactions can be complex and severe.

Cured epoxy that is softened with a heat gun is easy to remove with a palette knife. The edges of both parts of the joint were taped off before gluing to prevent the squeeze-out from getting on the work.

If you do get epoxy on your skin, don't wash it off with solvents such as lacquer thinner or acetone. They're health hazards as well, and will help your skin absorb the epoxy. Instead, use waterless hand cleaner or vinegar to remove epoxy while it's still wet.

This porous surface is being prewet with a coat of thin marine epoxy. Afterwards, a second, thickened coat will be applied to glue down the aluminum halves of a production routing jig.

Epoxy is more flammable than most woodworking glues. It can burn and explode, and should be handled like a finishing material. Exothermic batch mixes that overheat in the mix pot not only produce harmful fumes, but can also burn you on contact. Place overheated mixes in an empty metal pail and put the pail outdoors. The pail will act as a heat sink, and will contain any epoxy that cooks its way through the wall of the mix pot.

Performance

Epoxy is commonly regarded as the ultimate high-performance adhesive, and it does have some amazing performance properties. But it is a mistake to think that epoxy is a perfect, all-powerful miracle adhesive. It has some unusual performance-related requirements and also has a weakness or two among its properties. Since epoxy is often used for highly challenging gluing jobs, it's best to be as familiar as possible with all its performance-related characteristics.

Because epoxy is sold in many different forms and viscosities, you have to make sure that the epoxy you choose has the right application properties for the job at hand. You wouldn't want to use a thick, hard-to-spread paste for gluing up a bent lamination, for instance. In general, the epoxies that are used as wood adhesives are smooth and workable when properly mixed and spread easily because they have no tack. They can be rolled, brushed, or troweled without becoming overworked.

One reason why epoxy is a good adhesive is that it penetrates and wets bonding surfaces well once it's applied, especially the surfaces of highly porous and absorbent materials such as solid wood and sheet goods. In fact, in some cases you have to compensate for its penetrating ability as you apply it. Low-viscosity epoxies (such as marine epoxy) tend to overpenetrate porous surfaces, which leaves too little adhesive in the glue line to produce a strong glue joint. To prevent such starved joints, you have to prewet the bonding surfaces with a scant initial coating of epoxy. The prewet coat wicks into the bonding surfaces like a size and fills the pores so that a second, full application of epoxy will lie on top of the bonding surface. When you prewet your work, it's best to apply the second, full coating immediately afterward, before the prewet coat gels. The second coat should be thickened with silica or a similar additive so it will produce a glue line of ample substance.

WORKING TEMPERATURE AND MOISTURE
As epoxy formulas have evolved over the years, they have become less sensitive to the conditions they are used under. Older formulas require minimum working temperatures of 55°F to 60°F, but some newer epoxies will cure at temperatures as low as 28°F (even though they shouldn't be stored below

50°F). With multi-hardener epoxy systems, the faster hardeners work at lower temperatures than the slower hardeners. Likewise, fast-curing five-minute epoxies will work at lower temperatures than single-hardener, regular-speed epoxies.

Epoxy's sensitivity to moisture has also been improved. Older formulas don't work as well if either the humidity or the moisture content of the wood is too high. Newer formulas will work well even in high humidity and will bond wood that has a moisture content between 6% and 13%.

ASSEMBLY TIMES

As a group, epoxies offer a wide range of open and closed assembly times. Assembly times are determined by the same factors that govern pot life and by how soon you apply the epoxy after it is mixed. Applying epoxy as soon as possible after mixing it will give you longer assembly times. As you spread the mixed epoxy on your work, you disperse its volume, which prevents an exothermic reaction, slows its reaction rate, and extends its liquid life. If the work itself is cold (below 65°F), the assembly times will be lengthened even further. If you need substantially longer assembly times, you can switch to a slower hardener (if the epoxy system you're using offers one) or switch to an epoxy with a slower reaction rate.

CLAMPING PRESSURE

In most cases, epoxied assemblies require only light clamping pressure. The goal is simply to promote even, overall contact between the epoxy layer and the bonding surfaces. Heavy clamping pressure will drive too much epoxy out of joints and starve them.

GLUE-LINE THICKNESS

Epoxy has to cure in thicker layers than most other woodworking glues in order to develop the proper strength and endurance. Cured epoxy glue lines in wood should be thick enough to be seen with the naked eye. To produce such glue lines, use epoxy that has ample viscosity, apply it in the proper spread thickness, and only use light pressure to keep the epoxy from squeezing out excessively when the joint is clamped.

CURE PERIOD

Epoxy has two curing stages: the initial stage (gel stage) and the final stage. In the initial stage, the epoxy sets to a touch-dry state called "thin film set," but it is still fairly soft. It can be easily tooled and indented, and will move if a mechanical load is applied. At final cure, the epoxy is a hard solid that withstands mechanical loading. Simple, unstressed assemblies should remain clamped at least until the epoxy has reached initial cure, if not longer. Stressed assemblies such as bent laminations should remain clamped until the epoxy has reached final cure.

Cure speed is governed by the reaction rate of the epoxy mix and the temperature that the glued work is kept at. Epoxy's reaction rate slows down as the reaction progresses, so adding heat to a curing assembly will help shorten the cure period. It will also improve the ultimate properties of the cured epoxy. You can warm epoxied work to room temperature or above with a simple, low-tech heat setup such as the one shown in the photo on p. 112. Apply heat with extreme care and caution, as even the simplest devices can be hazardous if not used properly.

Epoxy is the best structural gap-filling adhesive available. Here, it compensates for an out-of-square corner block, while the beads of squeeze-out form fillets that strengthen the joint even further.

CURED WORKING QUALITIES

Once epoxy is cured, it can be filed and sanded, and some formulas can be hand-carved as in the bottom photo on p. 14. It machines well, and doesn't wear tool edges as much as urea resin or casein glues. It will resist stains if it infiltrates bare wood surfaces, and it may harm existing finishes if it seeps onto them. If trace amounts on bare wood are leveled and sanded back, the wood can usually be clear-coated without a problem. In fact, epoxy makes a good sealer, and will accept paints and other coatings once it's been sanded.

ENDURANCE QUALITIES

Epoxy is the most effective waterproof adhesive available. It is impervious to decay from organisms such as mold and resists many solvents. Most five-minute epoxies are not waterproof, but some new formulas have been developed that are.

Epoxy doesn't have great light resistance. Sunlight's ultraviolet rays may cause epoxy to degrade and lose some of its physical properties, which could lead to joint failure. This is primarily an outdoor concern, and not all joints are affected by it.

Most epoxies have fairly low heat resistance, and can be easily plasticized by heat. Epoxy's ultimate heat resistance is described as its HDT, or heat deflection temperature. The HDT of commonly used epoxies ranges from under 100°F to about 160°F or so.

If you warm your work to cure epoxy, monitor the squeeze-out. Adding heat can lower the viscosity of the epoxy, and it may flow out of clamped joints more readily. You may need to reduce clamping pressure to prevent excessive squeeze-out and joint starvation.

STRUCTURAL QUALITIES

Epoxy is the best structural adhesive available to small-shop woodworkers. Its combination of strength, toughness, rigidity, and shock resistance is superior to that of any other commonly used glue. Its performance exceeds what many woodworking jobs require. It's the only woodworking adhesive that can truly be called gap-filling on a structural basis. Even when heavily modified with additives, it still provides a superb bond.

There is some variation among different epoxy formulas, though. Standard epoxies tend to have a low initial strength and a high ultimate strength; fast-curing five-minute epoxies have a higher initial strength but a lower ultimate strength, which makes them less useful as structural adhesives.

8

Polyurethane Glue

Many people think of epoxy and polyurethane glue as companion adhesives. In a way, that makes sense. Polyurethane glue was originally developed as a substitute for epoxy and has some similar properties. However, in many ways, the two adhesives couldn't be more different.

Polyurethane glue was first made available to woodworkers in the United States in 1993, more than 20 years after its emergence in Europe. It arrived as a proven, refined industrial product that was introduced as a retail consumer glue. When woodworkers responded eagerly, large adhesive manufacturers noticed the growing market and introduced their own competing polyurethane formulas. Because the larger adhesive companies with the labs and research budgets were late entries into the polyurethane glue market, the volume of research results and technical literature that should be

available hasn't been produced yet. And if there is lots of European documentation on polyurethane glue, it isn't widely available.

Chemical Basis and Behavior

Polyurethane glue is not that different from polyurethane finishes. It contains resins that polymerize to form an adhesive film, just as the resins in polyurethane varnishes polymerize to form a coating. The glue's polymerization is driven by a di-isocyanate derivative base that reacts with water vapor, usually supplied by the moisture content of the material being glued and the ambient humidity of the shop atmosphere.

During polymerization, carbon dioxide is liberated as a by-product of the reaction. The CO_2 charges the wet glue with gas bubbles, turning it into an

Polyurethane glue got its start in Europe, and was introduced to woodworkers in the United States in 1993. It is now available in a range of viscosities and cure speeds.

Here's a case of foam-out. An undersized dowel has been glued into a drilled hole, and the polyurethane glue has expanded all around it and up onto the surface of the wood.

adhesive seltzer. The charged glue expands out of the glue line and reacts further on contact with the atmosphere, causing the oozing excess known as "foam-out." Foaming (see the photo below left) occurs only in places where the glue is unrestricted and allowed to expand, including gaps between gluing surfaces in sloppily made, poorly fit joints. In well-made, tightly clamped joints, the glue cures to a hard, durable, continuous solid, which is technically a polyurea, not a true urethane.

Formulation

Even though polyurethane glue is touted as a radical new miracle adhesive, it's simple stuff to a polymer chemist. The glue has only three major components, and there's not much a formulator can do to tweak its performance specifications other than vary the proportional relationships of these ingredients. Thus, there isn't a great deal of difference between the various glues on the market.

Most polyurethane glue is amber colored like its varnish cousin and dries to a light tan glue line. Some formulas are lighter colored and dry to a milky-white glue line, which isn't as visually pleasing in wood joints. Most glues have only a faint odor, and most have about the same molasses-like viscosity. Some newer formulas have a higher viscosity, all the way up to gel thickness. The solids content of polyurethane glue is typically very high, with some brands claiming 100% solids. Any solvent content is generally very low, from 0 to 5%. More important is the fact that the glue contains no water, so it won't cause any dimensional change in the glued work.

As a one-part, ready-to-use adhesive, polyurethane glue is designed to be more convenient to use than epoxy. This convenience is offset by the fact that epoxy is more versatile because it can be altered in various ways, as discussed in Chapter 7. Polyurethane glue can't be altered, but it does share epoxy's ability to bond a wide range of materials besides wood, such as metal and porcelain. Polyurethane glue also bonds many different species of wood well, including dense, oily, hard-to-glue woods. One if its most distinctive properties is its ability to bond woods that have a high moisture content (from 12% to 25%). However, wood that wet will change dimension as it dries further, which could cause glue joints to fail. Polyurethane glue won't work well with wood that's very dry. Its lower moisture content limit is about 8%.

This glue precured in the bottle when summer humidity migrated inside. Precure can be prevented by squeezing the air out of the container, capping it tightly, and storing it in a dry place.

Handling and Storage

Polyurethane glue is not as sensitive to storage temperatures as other glues are. It is freeze/thaw stable and can withstand summer heat extremes as well. It is sensitive to humidity, though, which can cause it to precure in the container. To maximize polyurethane shelf life, squeeze whatever air you can out of the bottle, cap it tightly, and don't open it any more often than you have to. Some glues skin over in the bottle just as polyurethane varnish does. If the glue under the skin appears to be still fluid, you can puncture the skin and transfer the glue into a smaller squeeze bottle for further use.

CLEAN-UP

After using polyurethane glue, the first thing to clean up is yourself. If you have any on your skin, remove it by the least toxic means possible. If the glue is still wet, use waterless hand cleaner. Don't use organic solvents such as naphtha, alcohol, or acetone unless you absolutely have to, as they will drive the glue into your skin. If you let the glue dry without cleaning it up, it turns dark and will remain on your skin for several days.

Clean gluing tools with the least toxic organic solvent that will do the job. As with epoxy, it's often easier to use cheap throwaway gluing tools than to clean up with solvents.

To clean up the glued work, wait until the glue cures and then pare off any excess. The spongy cured foam-out is a breeze to remove with either hand or

Cleaning up cured polyurethane glue is easy. Here, it's pared away with a sharp chisel, leaving a neat joint line.

power tools. Usually chisels and scrapers are the tools of choice. To clean wet glue from the work, you have to use solvents, and it's very hard to remove all traces of the wet glue from wood. It may also keep foaming out while you're trying to clean it up.

DISPOSAL

Before disposing of any polyurethane glue, let it cure to a foamy inert solid. An easy way to dispose of it is to put a shovelful of wood shavings in a plastic tub, moisten them with a plant sprayer, then dribble the glue over the shavings and leave it to cure overnight.

HEALTH AND SAFETY

Many people assume that polyurethane glue is harmless. However, the glue is a serious potential health hazard because its reactive base is an altered version of a highly toxic compound—methylene di-isocyanate (MDI). This compound is a skin, mucous-membrane, and respiratory-tract sensitizer. Regular skin contact can cause rashes, and repeated inhalation can produce reactions that range from a sore throat to intense asthma-like symptoms and pulmonary edema. Single exposures can cause headaches and dizziness. Isocyanates are also currently being tested for carcinogenicity. Although not everyone is susceptible to MDI sensitization, to be on the safe side you should wear disposable gloves and a respirator when working with polyurethane glue, and ventilate the shop during big glue-ups.

Performance

The simplest part of using polyurethane glue is grabbing the bottle and squeezing it out. After that, applying it to the work takes some care and effort. Most of the formulas are fairly viscous and hard to manipulate, especially in the scant spread thicknesses that are recommended. Because the glue expands as it cures, if you apply it as liberally as a PVA, it will foam out profusely.

Some suppliers recommend applying polyurethane glue to only one of two halves of a joint. I prefer to put the glue on both faces of wood joints. With its ample viscosity and lack of moisture or solvents, I'm not confident that the glue spread on one half of a joint will adequately coat the other half of the joint when the two are assembled. With polyurethane glue, there should be plenty of working time available to apply a thin but thorough coating to both parts of a joint (if you are veneering, though, apply the glue only to the substrate).

Like epoxy, polyurethane glue doesn't bond well to surfaces that are dead smooth. If your joints are hand planed, scuff the gluing surfaces with sandpaper before gluing up. Don't make the surfaces too rough, though, or the glue will foam slightly as it expands into surface irregularities, which will lower the strength of the cured joint.

ASSEMBLY TIMES

The open assembly times of most polyurethane glues range from 20 to 45 minutes, which will allow you to complete tricky, time-consuming glue-ups. Earlier polyurethane formulas are noticeably slower than newer formulas. The most reactive new glues are almost too fast to use for big glue-ups. Closed assembly times are also usually generous, and the glue has low initial tack, which makes parts easy to move around after assembly.

CLAMPING PRESSURE

Polyurethane glue works well only when the mating parts of a joint are well fit and clamped firmly to force the glue into a thin film, which will cure to a solid, consistent, and strong bond layer. If the glue is not forced into a thin film while it cures, it can foam up between the mating parts of the joint, creating a thicker and much weaker bond layer.

CLAMP TIME

Clamp times vary greatly, and again, the newer glues are more reactive than the earlier formulas, and cure more quickly. I leave the clamps on at least until the foam-out has cured to about the hardness of a Styrofoam coffee cup and can't be easily scraped off the work with a thumbnail.

To shorten clamp times, you can increase the reactivity of the glue and make it cure faster by lightly misting joint surfaces with water before applying

The curing speed of polyurethane glue can be accelerated by misting the gluing surfaces with water before applying the glue. Use only sparing amounts of water—too much moisture will impair the glue's reactivity.

the glue. Don't flood the work, because polyurethane glue has a water-vapor reactivity threshold—if too much moisture is present, it will act as a reaction inhibitor, not an accelerant. After the joint surfaces are moistened, let them dry to the touch on the surface before spreading the glue. Joints that have been moistened will probably foam out more than usual. If the moisture content of your work is below 8%, you need to moisten it to allow the glue to cure properly at all.

This is the dowel joint shown in the bottom photo on p. 86 after it has been sawn apart. The undersized dowel is encased by foamed glue, which has no structural strength. The joint half on the right was easily broken apart by hand.

CURING REQUIREMENTS

The minimum curing temperature of polyurethane glue is 50°F. It can be applied in temperatures colder than that if the work is warmed to above 50°F during the cure period. If you warm curing work, also add moisture to either the work or the surrounding atmosphere, because dry heat will retard, not accelerate, the curing speed of the glue.

STRUCTURAL QUALITIES

Overall, the ultimate properties of polyurethane glue are no match for those of epoxy. Nevertheless, polyurethane glue cures to a bond layer that's harder and more rigid than most PVAs, but less rigid and brittle than urea resin glues. These properties apply only where the glue has cured to a solid bond layer. Where polyurethane glue cures to a foam, it has virtually no durability or integrity. Thus, even though its expanding, foaming action does fill gaps, it doesn't fill them with structural strength.

CURED WORKING QUALITIES

Polyurethane glue has excellent overall working qualities. Besides being easy to work with hand tools, it machines smoothly without noticeably dulling cutting edges. Due to its friability, it's highly sandable, and it won't load up sandpaper. Once it's sanded back to a wood surface, it's stainable and is very compatible with most finishes.

ENDURANCE QUALITIES

Some suppliers claim that polyurethane glue is waterproof. Actually, it's highly water resistant, not waterproof, and the distinction is important. It will withstand intermittent exposure to almost any level of moisture, but will not withstand continual exposure to water at all.

One of polyurethane glue's outstanding properties is its heat resistance, which is much greater than that of epoxy or PVA. This makes it a good choice for things such as kitchenware and entrance doors that stand in direct sunlight. Polyurethane's good heat resistance and rigidity combine to give it excellent creep resistance.

9

Cyanoacrylate Glue

As a woodworker, I got along for years without cyanoacrylate glue—but only because of my own ignorance. I assumed that products such as cyanoacrylate glue belonged in a homeowner's utility drawer, not in a woodworking shop. I didn't take cyanoacrylate seriously because of the funny late-night TV ads touting its "miracle" properties and because of the way the glue was named and described: hot, super, instant, crazy, and so on.

Once I saw other woodworkers using cyanoacrylate and tried it myself, I quickly began making up for lost time. Cyanoacrylate (CA) glue is a great addition to the shop of any versatile woodworker. It was originally designed to bond non-wood materials such as metal and glass, but it also works very well with wood in certain applications. CA glue is best used for bonding small assemblies, rather than for typical woodworking glue-ups such as face

laminations. Because of this I don't use CA glue every day, as some guitarmakers and wood turners do, but when I do use it, it delivers fast, strong bonds, just as the ads claim.

Chemical Basis and Behavior

Cyanoacrylates are modified acrylic monomer compounds, which means that they are chemically related to various space-age plastics. In fact, you can think of these glues as liquid Plexiglas because they cure to a solid that's similar to common sheet acrylics. Like polyurethane glue, CA glues are moisture-curing. They polymerize through a chemical reaction that begins when they encounter trace amounts of moisture in a workpiece or in the surrounding atmosphere. Once they begin reacting, CA glues cure to a solid in a matter of seconds, like water

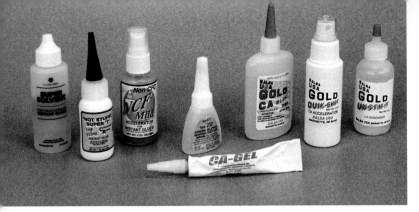

Cyanoacrylate glues will bond many different materials faster than any other adhesives woodworkers use. These glues come in various formulas and viscosities, and in consumer and industrial grades.

droplets flicked onto a frozen metal plate. (If the glue overreacts, the effect becomes more like water drops bubbling on a hot skillet.)

Though CA glues are advertised as all-purpose adhesives, not all CAs will bond all types of materials. Most CAs are sensitive to the nature and condition of the materials being glued. For example, they won't cure if the pH of the bonding surface is too low (acidic), and they cure more rapidly than usual if the pH of the bonding surface is too high (alkaline).

Formulation

Like other adhesives, cyanoacrylate glues are manufactured for industry and consumer use. The glues that most woodworkers use are the consumer ("hobby") glues. Industrial CA product lines are extensive; they include dozens of different formulas, each designed for a specific job, such as bonding rubber or spot-gluing electrical components. Consumer CA product lines are limited to five or six different glues at most.

Cyanoacrylate glues are formulated from a handful of different monomer resin bases, such as ethyl, methyl, and propyl. These bases determine the working properties of the glue and its suitability to different tasks. To keep retailing simple, consumer CA glues are sold on the basis of physical properties such as cure speed and viscosity. For most woodshop gluing tasks, choosing glue on that basis is fine. For specialized gluing tasks, you may want to choose an industrial-grade glue after reviewing product information that specifies monomer-base types along with other properties and references glues to the tasks they're designed for.

PURITY

The quality of a cyanoacrylate glue depends on the purity of its monomer base resin. Monomer base resins are purified through distillation, and their purity is determined by how carefully and thoroughly they are distilled. The higher a glue's purity, the higher its ultimate properties, such as tensile strength, will be. Highly refined glues also bond a wider range of materials with greater compatibility than glues of average purity. In general, you can assume that industrial-grade CA glues are highly refined. With hobby-grade glues, you may want to check the purity rating if one is available. For best results, buy glue whose purity rating is as close to 100% as possible.

COLOR

Most consumer CA glues are semi-clear and colorless. Industrial product lines include high-clarity formulas and solid-color glues that are made with additives such as black rubber. Avoid CAs that are yellow or amber because they may have been overheated while being distilled or may have excess added stabilizer.

ODOR

Many CA glues give off noxious fumes that can irritate your eyes and nostrils. These fumes can linger and build up in your shop. Fortunately, new "low-odor" and "odor-free" formulas are now available, and these are far less irritating to use than the regular formulas. If you use these new glues, test them first to make sure that they will bond your materials properly, and use them as carefully as you would use a noxious glue.

VISCOSITY

Consumer-grade CA glues are sold in several viscosities, from a watery syrup to a thick gel. Thin glues wick into confined spaces very well and are used for things like fixing cracks in glass and porcelain. Thicker glues are used for gluing uneven surfaces together and for filling larger cracks, such as surface checks in solid wood.

Often, thin and thick glues are just different versions of the same basic formula, and they have many similar properties. If you find that the glues you work with aren't versatile enough, you have two options: Use a more highly refined glue, which can be used with a wider variety of materials, or buy an industrial glue that is specifically formulated for the work you're doing.

COMPONENTS

Along with their monomer resin bases, cyanoacrylate glues contain two other key components—thickeners and stabilizers. Thickeners allow a

manufacturer to produce glue in different viscosities. Stabilizers are used to preserve and extend shelf life and to compensate for impurities in the glue. Most cyanoacrylates have some percentage of added stabilizers. Less refined glues contain a much higher level of stabilizers—and are known as "up-stabilized"—but they have much lower cured strength and durability.

Many CA glues are sold as systems that have other components besides those that are blended into the glue itself. Both hobby-grade and industrial CA systems include an accelerator, which is a catalyst that is used when the fastest possible bonds are required, and a debonder, which is used to disassemble parts that have been mistakenly or incorrectly glued together. Industrial systems also include other components such as surface prep agents, primers, and adhesion promoters, which are used to maximize the quality of the bonds that CA glues will produce.

Cyanoacrylate accelerators consist of a small amount of an alkaline compound, such as an aromatic amine, carried in a solvent medium. The accelerator's alkalinity increases both the rate and the intensity of the glue's reaction and cures it to a solid mass in the blink of an eye. Some accelerators are carried in aggressive solvents such as acetone, which will soften various bonding materials such as plastics and rubber and will soften and harm wood finishes. When using CA glue for repairs on finished work, I avoid acetone-based accelerator and use a milder product such as a heptane-based accelerator instead. For more on using accelerators, see the sidebar on p. 94.

Using Accelerator

If you're going to use an accelerator to speed-cure CA glue bonds, you can apply it either before or after you assemble the glued work, depending on the job at hand and your preferred methods of working. To apply accelerator beforehand, apply glue to one half of the joint, and the accelerator to the other half.

The accelerator will catalyze the glue when the two halves are assembled. Your other option is to apply accelerator to the joint once it is assembled. This is somewhat less desirable because the accelerator has to harden the glue from the outside of the joint inward and may not migrate far enough to reach glue buried deep in a joint that has a lot of surface area. All the glue in the joint will cure eventually, though.

In general, don't use an accelerator if you don't have to because bonds cured with accelerator are much weaker than unaccelerated bonds and have less optical clarity, which is important when gluing a material such as glass. If you do use accelerator, use as little as possible to avoid frosting the bond. Frost, also called chlorosis, occurs when the glue overreacts as it cures. It foams and turns white, and emits more fumes than usual. Frosted bonds are weaker than carefully accelerated bonds.

If you don't want to use an accelerator to speed the cure rate of CAs, you can simply use moisture instead. You don't need much moisture to make things happen—sometimes, just breathing on a workpiece before applying glue will add enough water vapor to the bonding surfaces to promote faster curing.

Accelerator makes things happen fast, and it's easy to use it habitually. Don't use it if you don't need it, though. Unaccelerated bonds are stronger and produce less noxious fumes.

Breathing on work (such as this tool handle and metal ferrule) just before applying cyanoacrylate glue adds enough water vapor to the bonding surfaces to accelerate the moisture-curing glue. For a faster bond, breathe more heavily.

Debonders are based on solvents such as nitromethane and acetone that will dissolve the glue, even after it has cured to a solid. Debonders will also break down some bonding materials and wood finishes, so use them carefully to avoid damaging glued work.

SURFACE SENSITIVITY

Many CA glues are sensitive to the pH of joint surfaces, and their surface sensitivity is a matter of degree. Highly acidic surfaces will impair or prevent bonding, but slightly acidic surfaces may only retard the cure time of a glue.

Similarly, slightly alkaline surfaces bond readily, but a highly alkaline surface can cause the glue to overreact and "shock bond," which reduces the available working time and results in a weaker bond. If you're not sure whether the materials you're gluing are acidic, alkaline, or neutral, you can test your glue on small sample or scrap pieces. If problems occur, contact the supplier or manufacture of the glue and/or the materials. You can also consult technical literature in engineering school libraries and other locations, or order materials conservation literature from sources such as the American Institute for Conservation (1717 K St. NW, Suite 301, Washington, DC 20006) or the Smithsonian Institution Museum Support Center (4210 Silver Hill Rd., Suitland, MD 20746).

When you use cyanoacrylates that are surface sensitive, you may have to prepare various bonding surfaces before gluing them. If the bonding surfaces are highly alkaline, you can keep them from causing the glue to overreact and shock-bond by swabbing them with a solution that's slightly acidic, such as a highly diluted mixture of white vinegar and distilled water. A more convenient solution is saliva. Let the bonding surfaces dry before gluing them.

The pH of most bonding surfaces is easy to neutralize if needed. Alkaline surfaces can be swabbed with a simple, mildly acidic solution such as white vinegar and distilled water (saliva is a convenient alternative), and dried before gluing. Acidic surfaces can be swabbed with solutions made from alkaline materials, such as the ones shown in the top photo on p. 96. Cyanoacrylate accelerator is the most convenient of these to use. Again, dry the surfaces

You can use debonder to free two objects bonded with cyanoacrylate, in this case, a bicycle helmet that was inadvertently glued to a metal beam.

before gluing them. Test all surface preparations before using them on actual work, because acidic and alkaline solutions can attack finishes and change the color of bare wood.

Cyanoacrylates are also very sensitive to surface contaminants such as grease, residues, and rust. Contaminants have to be completely removed for the glue to bond properly. The porosity of the materials being glued affects bonding as well. Cyanoacrylate glues are designed primarily to bond non-porous, non-wood materials; wood that's very porous can be hard to bond with CAs. Low-viscosity glues wick into the pores before the glue can cure, which leaves the glue line starved and prone to failure. For porous woods, many woodworkers use either a medium- or high-viscosity glue to prevent starvation.

Cyanoacrylate glue is surface sensitive and might not bond these cherry jig parts properly unless their natural acidity is neutralized with a mild alkaline surface rinse such as those shown.

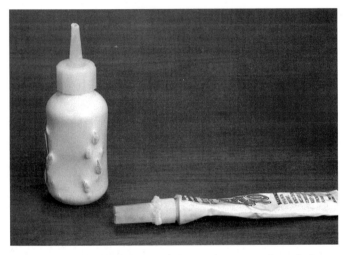

Keep cyanoacrylate glues dry and store them away from their accelerator or any other alkaline materials, so they don't precure in their containers before their shelf life expires, as these glues did.

Handling and Storage

In most shops, cyanoacrylate glues are used less frequently and in smaller amounts than other adhesives. Consequently, woodworkers may not pay much attention to storing and handling them. Over the years, though, I have found that CA glues actually require more careful storage and handling than other glues. Even if you use only a drop or two of CA glue once or twice a year, you should treat the stuff with care because it's powerful and can be hazardous.

SHELF LIFE

Most CA glues have a claimed shelf life of about six months. Thin glues are more reactive than thick glues, and have shorter shelf lives. Monitoring shelf life is usually easy because properly stored CA glues of all viscosities gradually thicken with age. To maximize shelf life, buy glue at its freshest from suppliers who replace their stock frequently, avoid the dusty bottles on the back shelf at the hobby shop. The labels of some glues are dated so you can check their freshness. When you buy glue, mark the date on the container if it isn't dated, then refrigerate the glue unopened. Let it warm up to room temperature before opening the bottle and using it for the first time.

Once the container has been opened you shouldn't refrigerate it, because condensation may form on the inside of the bottle and precure the glue. But you can still prolong shelf life by storing glue bottles in airtight containers or in boxes with desiccant packets. Don't store glue together with its accelerator, or with any other alkaline material, such as lye, bleach, ammonia or epoxy hardener, because these may off-gas vapors that will precure the stored glue. Keep the tips capped to prevent moisture from migrating into the bottle.

Containers and Applicators

The usefulness of a cyanoacrylate glue depends on the way it's packaged. The greatest formula in the world won't do you any good if it's sold in a poorly designed bottle. A good bottle keeps moisture out and is easy to squeeze. The top should be easy to remove from the bottle for cleaning; it should also resist clogging and allow glue to be applied precisely in small amounts. The cap should seal properly and resist becoming bonded to the tip. You can help keep tips and caps in good shape by working neatly and by tapping glue back down into the bottle after you've squeezed out what you need. This amount of fuss may seem excessive, but these details are important if you depend on CAs to be both speedy and convenient to use.

Packaging also affects the usefulness of the other CA system components. Some accelerators are sold with squirt pumps that spray the fluid crudely and uncontrollably. Some debonders can't be applied by the drop; they stream out onto glued work, which can be disastrous.

I buy extra tips and caps and also save them from emptied bottles. I also save old applicator brushes, pump sprayers, and debonder bottles if they work well. With these extra parts on hand, if a tip becomes clogged, I have a clean one to put on the bottle while I soak the clogged one in acetone or debonder. If I don't want to spray accelerator on a workpiece, I pour it into a reclaimed debonder bottle and drip or brush it on.

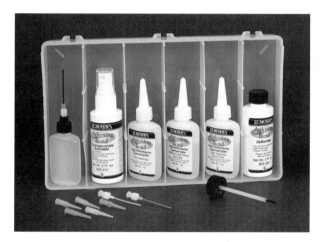

Cyanoacrylate glues are effective and easy to use only when they are packaged in good containers. This handy system package includes glues, debonder, accelerator, and several applicator tips.

You can soak clogged tips and caps in acetone to clean out cured glue. Keep a few spare clean tips and caps on hand so you can replace any clogged tips and caps that need cleaning.

CLEAN-UP

There usually isn't a lot of clean-up to do with cyanoacrylate glue because it's applied precisely in small amounts. But if you need to wipe up excess CA, use debonder or acetone as a cleaner. Work as quickly and carefully as you can because the glue is easier to clean up when it's still fluid.

DISPOSAL

Cyanoacrylate glue can be safely and easily disposed of because it cures to an inert solid, which is a benign waste product. Don't discard any glue that hasn't completely solidified, though. The easiest way to solidify CA glue is to let it cure in the bottle. If you're in a hurry to dispose of it, unscrew the top of the bottle and place it in an empty covered metal container (such as a safety waste can) together with an open bottle of accelerator or a jar of strong baking-soda solution, which will harden the glue within a matter of days. Don't cure CA glue for disposal by pouring accelerator or other alkaline compounds into the bottle, because they will react violently with the glue inside such a small container. Don't remove glue from the bottle and accelerate it inside the shop because the reaction may generate lots of fumes. Also, don't pour the glue into a pile of moist shavings or a wet rag, because it can become exothermic if it cures in contact with lots of moisture.

Working Safely with Cyanoacrylate Glues

Here are some ways to make working with CA glues as pleasant and safe as possible:

• Work in a well-ventilated area.

• Use a low-odor or odor-free glue after making sure it suits the work at hand.

• Don't use accelerator unless you need to. If you need an accelerator, try a safe, non-toxic alternative such as baking soda.

• If you need to use a debonder, use the least toxic one available.

• Wear gloves and a respirator.

• Wear lab goggles instead of safety glasses.

HEALTH AND SAFETY

As you work with CA glues, it's hard to ignore the hazards they present. Most noticeably, they give off noxious fumes that irritate the eyes and the upper respiratory tract. These fumes are not highly irritating when you use the glue occasionally and a drop at a time, but they can become overwhelming if any measurable volume of glue is used. Even a dram of the stuff can chase you out of a poorly ventilated 500-sq.-ft. work area. The fumes become worse if you use an accelerator, and they can linger in the shop long after the glue has hardened.

Cyanoacrylates are notorious for bonding human skin. It's easy to glue your fingers together or to glue your hand to a tool handle or a workpiece. Bonded skin can be unglued with debonder or solvent—provided you have a free hand and can open the container.

Cyanoacrylates can also cause skin conditions such as contact dermatitis. These conditions are caused more by exposure to glue vapors than by skin contact. CAs can also cause occupational asthma and are currently being tested to determine if they are potential mutagenic and carcinogenic agents.

Cyanoacrylate glue can permanently damage your eyesight if it gets in your eyes, which is more likely to happen than you might think. I once had some thin glue squirt wildly out of a bottle whose tip I was unclogging, narrowly missing my face. Protect yourself from these hazards by wearing disposable gloves, goggles, and a respirator.

Performance

Despite being touted as "miracle" adhesives, cyanoacrylate glues are not very forgiving. They either bond materials together very well, or they don't work at all. To ensure good performance, you have to use them as they were designed to be used. This means making sure that the glue you're using is well matched to the task at hand and to the materials being glued. It also means making sure that the gluing surfaces of those materials are prepared if needed to promote effective bonding.

APPLICATION

Cyanoacrylate glues work best when applied in sparing amounts and when pressed to a thin bond layer. Apply glue to only one of the two bonding surfaces, spread it to an even coating if needed, and close the joint. Palette knives and scraps of sheet plastic cut from milk jugs make great spreaders. If you're using thin CA glue, be careful to avoid runs and drips, and if you're bonding porous surfaces, check to make sure that the glue hasn't wicked into the pores and left the bonding surfaces starved. A good way to prevent starvation is to prewet porous gluing surfaces with thin glue, then apply a coat of thicker CA glue just before closing the assembly.

ASSEMBLY TIMES

Even though CA glues bond in seconds, they often allow open assembly times that are a bit more generous. When you apply CA glue to a clean, dry bonding surface, you may have over a minute to spread the glue before it starts hardening. Higher-viscosity glues allow more open time than thin formulas. Some woodworkers use thicker glue just to get a few seconds more open time.

You can extend both the open and closed assembly times by gluing up in a shop that's colder than 50°F. Cold temperatures retard the reaction rate of the glue. Some glues are so sluggish when cold that you may have to use an accelerator to get them to cure.

Once you close a CA-glued joint, you have scant seconds to position the workpieces before the glue starts to solidify. As is the case with open time, thicker glues have a longer closed assembly time than thin glues do. Regardless of your glue's viscosity, if you apply more glue to the work, you'll have a longer closed assembly period. Nevertheless, in many cases, you won't have time to put clamps on the work. Use hand pressure to close the assembly, and hold it until the glue hardens.

Cyanoacrylate glue mixed with sawdust makes a good wood filler for cracks, holes, and other voids. The glue cures in seconds and can be sanded as soon as it begins to harden.

STRUCTURAL QUALITIES

Cured CA glues have good rigidity, ample tensile strength, and adequate shear strength for most small-shop applications. Odor-free glues generally have a bit more cured strength than regular glues. However, CAs have low shock resistance, and odor-free glues are far less shock resistant than regular glues. You can shock a CA-glued wooden assembly apart with a tap from a small cabinetmaker's hammer. In my experience, though, the durability of the glue varies, depending on the nature of the materials being bonded. I can make very durable joints with CA glue in pliant materials such as sheet rubber, which absorb shock loads instead of transferring them to the glue joint. A CA bond's ultimate strength is achieved within 24 hours after it hardens.

GAP-FILLING ABILITY

If there are any gaps between the bonding surfaces, thin glue will tend to run out of the gaps, but thicker glue will fill them. Either way, the glue takes longer to cure when there are gaps in the work. Cyanoacrylate glue is one of the few glues that is a true structural gap filler, although filled gaps aren't as durable as a thin, even bond layer.

CURED WORKING QUALITIES

Cured CA glue can be filed, ground, and sanded easily without loading up tools or sandpaper. It will resist wood stains and finishes, so it has to be carefully removed from primary visible surfaces before finishing. Most CAs have good resistance to moisture and chemicals (other than their own solvents). They also have superb resistance to low temperatures, but low resistance to heat and only fair resistance to ultraviolet light.

10

Using Glue Successfully

Having looked at individual adhesives and their properties in the previous chapters, it's now time to turn our attention to the use of those adhesives in the workshop. In this chapter, I'll focus on many factors other than the glue itself that can affect gluing and the results you get from the process, and I'll discuss various things that you can do to control these factors and improve or perfect the way you use glue in your shop. This information is organized into four sections that cover the full scope of the gluing process: preparing your shop, preparing your materials for gluing, gluing up, and conditioning your glued work.

Preparing Your Shop

If you want to achieve consistently great gluing results, the first thing to do is make sure that your shop is as well suited to gluing as possible. I've learned firsthand over the years that poor shop conditions make gluing much more difficult and affect the quality of glued work. Fortunately, the conditions in most small shops can be significantly improved just by addressing a few basic issues and correcting shortcomings with simple, direct solutions.

TEMPERATURE

The first step in improving the conditions in your shop is to gain as much control over its temperature as possible. The more control you have, the better your gluing results will be. Ideally, your shop temperature should be between 65°F and 80°F. However, keeping a shop within that temperature range all year is difficult for many woodworkers. If the temperature in your shop is hard to control, you can compensate for excessive heat or cold by using slower-acting glues and working

Shop owners once went to great lengths for a good glue-up. The bank of steam pipes along the rear wall of the room is used to heat both the hide glue and the workpieces before the glue-up. (Photo © Norman Hurst, courtesy The Old Schwamb Mill.)

Successful glue-ups depend on a lot more than glue. This finely dovetailed hide-glued drawer has endured 150 years of use because it was skillfully made from well-chosen, well-seasoned lumber.

faster in summer (when most glues harden more quickly) and warming your work before and after gluing in winter (when most glues have a difficult time hardening properly).

Whenever possible, try to use your shop's temperature to your advantage. For example, when I use hide glue in a cool shop, it gels very quickly, which allows me to glue up small, simple assemblies at a rapid pace. On the other hand, glues like urea resin glue harden more slowly in a cool shop, allowing extra time to glue up complex assemblies. In a hot shop, the advantages are reversed. Hide glue is easier to use for complex glue-ups because it takes longer to gel, and assemblies glued with urea resin glue can be unclamped sooner because the glue hardens more quickly.

HUMIDITY

The humidity of your shop is worth controlling because it affects how your glues perform. Some adhesives (e.g., epoxy) don't work well in high humidity, yet others (e.g., polyurethane glue) thrive on it. I try to keep the humidity in

Water quality is important to gluing. Even a cheap faucet-mounted filter unit like this one can be a big help. The filter cartridge on the left is new; the one next to it is several months old.

Excessive air movement makes gluing more difficult. This ceiling-mounted heater is fitted with a metal vane to help control shop airflow.

my shop at 50% year round with a humidifier and a dehumidifier so that I can use any glue at any time with good results. Controlling humidity can also help you apply finishes properly and control the condition of lumber and other materials.

LIGHT

You may need to increase or decrease the amount of light in your shop. If your workspace is flooded with direct sunlight, glue joints that remain exposed to ultraviolet rays for long periods may degrade or break down. If your shop is too dark, you may not be able to see well enough to apply glue properly or to remove excess glue from your work as thoroughly as you should.

AIRFLOW

Although ventilation is important when using glue, lots of active airflow can make glues such as PVA dry too quickly. My ceiling-mounted, gas-fired shop heater warms my shop fast in winter but also pushes a lot of air around, so I turn it off during big glue-ups. I also fitted it with an external fin to direct some of the air away from areas where gluing takes place (see the photo above left). In summer I shut off all fans and close windows on windy days before gluing up.

WATER QUALITY

For years, I mixed powdered glues with any water that was within reach. Then I learned how important water quality can be to glue mixtures. Water can carry oil, grease, and excess minerals, and may be either acidic or alkaline. Glue batches made with poor-quality water can be hard to mix. They may contaminate gluing surfaces inside the glue joint, or they may cure to inferior glue lines. Use bottled water to mix glue, or filter your tap water if it's suspect. Use distilled water for rinsing metals and other materials whose gluing surfaces require careful preparation.

You should also use good-quality water to clean up excess glue from your work. Poor-quality water can migrate into joints as you clean up around them and contaminate part of the glue line or stain the surface of bare wood around joints that are being cleaned up.

GLUE STORAGE

How you store glue determines its shelf life and general usefulness. Storing glue is not a big deal: All you have to do is keep it away from temperature extremes, high humidity, and light. If you are having trouble controlling conditions in your shop as a whole, you may not be able to store glue properly by simply leaving it in a cabinet, but you can set up various simple arrangements. You can refrigerate some glues in hot weather. In winter, you can keep water-based glues in a cabinet fitted with a small light bulb (wired carefully and mounted safely inside the cabinet using the proper materials). The heat of the light bulb will keep the glues from freezing. Glues can be isolated from humidity in sealed plastic bags and should be kept out of direct sunlight.

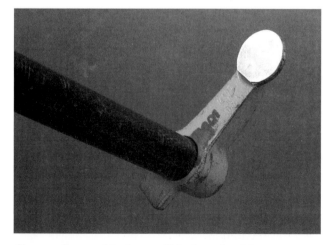

Clamps often need tuning up. This clamp head has a permanent vinyl pad, and its casting numbers have been ground off so they won't dent clamped work.

SAFETY

Many people don't think safety is a big concern with glues, but it is. Most of the synthetic adhesives woodworkers use are hazardous in one way or another, and often the more powerful and effective a glue is, the more hazardous it is to your health. Your shop should have some means of ventilation or air exhaust, a first-aid station, safety gear such as respirators and gloves, and goggles as well as safety glasses. Keep an empty metal pail handy as a safety container for overheated batch mixes of epoxy.

TOOLS AND MACHINERY

To make good glue joints, your tools and machines have to be in good working order. Dull cutting edges can burn, rough up, or burnish gluing surfaces, and improperly installed jointer or planer knives will produce wavy, scalloped surfaces that won't glue well. Machine surfaces that have a heavy buildup of grime, wax, or lubricants can transfer these contaminants to gluing surfaces as you machine them.

You should keep your glue applicators, clamps, and other gluing gear in good shape, but there's no need to spend top dollar for fancy equipment unless you want to. I usually buy modestly priced equipment and tune it up to maximize its performance. For instance, I make pipe clamps from the highest-quality heavy black steam pipe, which I clean and wax after it's threaded. I true up clamp faces on a disc or belt sander if I need to and glue permanent vinyl pads to them with cyanoacrylate, polyurethane glue, or epoxy. I also soak new wooden handscrews in naphtha to remove their nondrying oil finish so they won't stain my work, and I coat them with gel varnish so glue won't stick to them. It's worth paying this kind of attention to all your clamps.

Clamping glued work can often take as much room as a small shop can spare. To make the most of limited shop space (and to make clamping easier if you're working by yourself), you can rig up strategic devices such as clamping frames, or sawhorses with notched beams. Fixtures like these hold your clamps in perfect position for edge-gluing and frame assembly and can make glue-ups much easier. You can also make good use of shop fixtures you already have. I set up the vacuum press on the table saw's catch table so that I can load veneered work easily into the bag by sliding it across the saw table.

Preparing Your Materials

The more carefully you store, handle, and prepare the materials you glue together, the better your glued work will be. This doesn't mean that you have to fuss constantly over your lumber, sheet goods, and veneers, though. Simply store them sensibly, join them as skillfully as you can, and make sure their gluing surfaces are in good shape before you bond them together. While you're getting materials ready to glue, try to pay attention to their basic physical characteristics, such as flatness, sponginess, or brittleness, so you can take advantage of or compensate for these features as needed when you glue up.

DEALING WITH THE PROPERTIES OF WOOD

In contrast to glues, which are carefully formulated uniform products, wood is a slightly refined, natural raw material. When you glue wood, the glue is a constant factor, and the wood is a variable factor. To produce glued work of consistent quality, you have to take into account the variability of the wood you're gluing.

Wood varies widely, not only from species to species but also from plank to plank within the same species, depending on where the tree grew, how it was harvested, how it was sawn into lumber and dried, and how the dried lumber was stored. Given this variability, there will likely be imbalances of some sort between any two pieces of wood you glue together. For example, if you glue two pieces of wood together that have different densities and porosities, the glue will wet and penetrate them to different degrees and may adhere better to one piece than the other. Similarly, veneer varies from species to species and from flitch to flitch, and you'll find that the same spread thickness of glue will bleed through different slices of veneer to different degrees.

The most important characteristic to monitor is moisture content. Two pieces of wood being glued together should have the same moisture content, and they should both be at a point of equilibrium with the surrounding environment. It's not enough for two pieces of wood both to be at 15% moisture content at glue-up, because they are both going to keep losing moisture until they reach an

About Creep

Creep (also known as cold flow or cold creep) is a common problem in woodworking. When a glue joint creeps, its parts shift into a slight but perceptible misalignment after the glue has cured and the joint has been leveled flush (see the drawing below right). Some woodworkers think that glue causes creep by shifting and moving on its own, bringing glued parts along with it. Actually, wood causes creep and glue lets it happen. Wood can impart a great deal of mechanical force to a glue joint through factors such as hygroscopic movement, moisture-content imbalances between glued parts, and stresses within individual parts. A cured glue line has to be very rigid to resist these forces and hold the work flush. Some glues (such as consumer-grade PVAs) simply don't cure rigidly enough to keep parts from changing dimension or alignment.

Heat is also a contributing factor. Some thermoplastic glues begin to soften at temperatures as low as 110°F, which greatly lowers their creep resistance regardless of how rigid and creep resistant they are at lower temperatures.

COLD CREEP

Drier board — Wetter board

Two boards with different moisture contents were edge glued, then planed flush. As the boards released moisture, the wetter board shrank more in thickness, and the glue layer couldn't prevent the joint from creeping out of flush alignment.

equilibrium point of 8% or so in an indoor workshop. The shrinkage that will occur can easily wreck the glue joint (see the sidebar above).

If you have a moisture meter, you can check the moisture content of your wood at various stages of a project prior to gluing up. If the moisture content is too high overall or varies from one workpiece to the next, stack and sticker project parts in the warmest, driest part of the shop so they can dry and reach equilibrium. Check them with the meter every few days until they are ready to glue. If you don't have a moisture meter, simply leave the parts stacked and stickered for as long as possible.

PREPARING JOINTS
The quality of the joints you make is more important to the endurance of your work than the type of glue you use. If you make great joints, any one of several glues will likely hold them together for many years, but if you make sloppy joints, most glues won't be able to hold them together for very long.

To start with, joints have to be well designed so that mechanical and seasonal hygroscopic stresses are avoided or minimized. They also have to go together neatly and tightly without excessive clamping pressure, which can add unwanted stress to the work. And they have to allow room for glue between the joint surfaces, but not so much room that the glue cures in thick bond layers, which are less durable.

You can make great joints more easily and consistently if you're equally adept with hand tools and machines. Although some joints can be machine-cut with great precision, other joints may need to be fine-tuned with hand tools to improve their fit. Sometimes, just skimming off a delicate shaving or two with a paring chisel or a plane can transform an adequate joint into a superb one.

Besides fitting well, joint surfaces also need to be in good condition. Most important, they have to be clean. Any wax, oil, grease, grime, or other contaminants that are present will keep glue from bonding properly. Joint surfaces can be wiped with solvents (such as alcohol or acetone), lightly scraped, or scuffed to remove any contaminants. Joints made of dense, oily, hard-to-glue woods can also be solvent wiped, or they can be machined or hand tooled just before glue-up to remove any surface oils and to expose fresh surfaces that glue can penetrate more easily. In general, try to glue your work together soon after it's prepared. The longer that prepared gluing surfaces remain exposed to the open air, the less likely they are to bond properly.

While gluing surfaces shouldn't be crudely worked, they don't always have to be perfectly smooth either. A carefully

A moisture meter is useful in many aspects of gluing. Here, two boards are checked for comparable moisture content before being edge glued.

bandsawn tenon can be glued just as effectively as a tablesawn or a hand-planed tenon. Some adhesives, such as epoxy and polyurethane glue, actually work better on surfaces that are not dead smooth. To use those glues with joints that have been made smooth, such as edge joints, scuff the gluing surfaces lightly with sandpaper first.

Gluing Up

As with improving your shop and preparing your materials, it's important to refine your approach to gluing up as much as you can. A good way to do this is to divide the gluing into its four main stages—preparation, glue application, assembly and clamping, and clean-up—then fine-tune each as needed.

PREPARING TO GLUE UP

It's important to prepare thoroughly and diligently for glue-ups because there can be lots of details to tend to in any glue-up, no matter how large or small it is. Even a seemingly minor item like an extra piece of masking tape can noticeably affect the success of a glue-up. I typically spend far more time getting everything ready for a glue-up than I do actually spreading glue and tightening clamps. Working this way removes the anxiety and uncertainty from gluing up and puts me in complete control of the procedure. To me, an orderly, efficient glue-up that proceeds like clockwork is a great payoff for the effort I've invested in preparing for it.

In my experience, the best way to prepare for a glue-up is to run through it without using glue, which is called dry-clamping. Many woodworkers dry-clamp their work before gluing it up, but often they do so just to make sure their joints are tight. I like to run through a glue-up in complete detail because it allows me to determine everything that's needed on a practical, firsthand basis. Every time I start a glue-up without making sure that everything is ready beforehand, I find that I've overlooked something and have to scramble to make up for it before the glue dries.

APPLYING GLUE

One of the most elusive skills in woodworking is applying just the right amount of glue. If you apply too little,

Handling Big Glue-Ups by Yourself

Big, complex glue-ups pose a great challenge to those working alone in a small shop. Often there's no turning back once you begin spreading glue. All the labor you put into a project up to that point could be lost if the glue-up fails. Yet glue-ups have to proceed, regardless of whether help is available or not.

There are many things you can do to make big solo glue-ups less frantic and risky. Improve your shop for better gluing in general as discussed earlier in the chapter, adding any aids and fixtures you can that will expedite your gluing. Design big projects so that they can be assembled in a sequence of manageable glue-ups rather than one giant glue-up. If big assemblies are unavoidable, try to glue them up in stages if possible. For instance, you can glue up a big carcase one or two corners at a time, assembling and clamping the whole carcase with each glue-up to keep all the parts in square alignment.

Before you glue up, choose the best glues and applicators for the job. Slower-acting adhesives that offer longer assembly times are ideal for large glue-ups. Applicators should allow you to coat the work thoroughly and quickly. Use inexpensive applicators so you can station several of them around a big glue-up as needed. Finally, after you rehearse the glue-up, get the shop ready for the actual procedure. Lower the shop temperature if possible to extend assembly times (raise it if you're using hot hide glue or hot melt), eliminate any airflow, unplug the phone, and lock the door.

you starve the joint; if you apply too much, it runs out of the joint and makes a mess. No matter how much experience you've had applying glue, you can apply the wrong amount just as easily as the right amount, even if you've been gluing all day. Though you should apply glue in careful amounts, don't be bashful about where you put it. I put glue anywhere in a joint where it has a chance to help hold the work together, including all the end-grain areas, even though they're not optimal gluing surfaces.

You may want to buy or make some glue applicators so you can regulate the amount of glue you apply to gluing surfaces such as solid-plank edges and veneer substrates. Whether your applicators are simple or fancy, keep them in good shape so you can apply glue efficiently. Don't chew the tips of clogged squeeze bottles to loosen and remove caked, dried glue; this enlarges and deforms the spout holes. Instead, soak bottle tips clean when they get clogged, and keep extra replacement tips on hand. Keep pumps, rollers, brushes, squeegees, and other equipment clean as well.

Spread thickness Adhesive tech data sheets (see the sidebar on p. 8) often include recommended spread-thickness specifications. Spread thickness is usually gauged in pounds per thousand square feet of single glue line, or lb./MSGL. This measurement is used in industrial operations, where glue spreading and clamping are done with machines. In a small shop, where glue spreading and clamping are done by hand, MSGL figures are best used as starting reference points to help you determine the spread thicknesses you

Glue-ups require organization and complete concentration. Here, joints are aligned for glue spreading and assembly, with caul blocks ready at hand. (Photo by Judith P. Hanson, courtesy B. A. Harrington.)

Glue applicators don't have to be fancy, just well chosen and well cared for, no matter how little they cost. Some of these are several years old.

need for your work. Using a formula published by *Fine Woodworking* magazine contributing editor R. Bruce Hoadley in his book *Understanding Wood* (The Taunton Press, 1980), you can convert MSGL specifications to figures that are more useful for small-shop gluing by dividing the given amount in pounds by 2.2 to get grams per square foot of surface area. For example, a spread thickness of 55 lb./MSGL can be converted to 25g per sq. ft. Most small scales read in both grams and ounces, which makes gram measurements easy to use.

Controlling spread thickness is most important when you're covering large surface areas, as in veneering. To regulate a spread using MSGL figures, calculate the total square footage of the substrate you're veneering. Then multiply the square footage by your glue's spread-thickness specification (after converting it to grams per square foot). This will give you the total weight of the glue you'll need to coat the substrate. Weigh out the glue, apply all of it to the substrate, and note the film

thickness. Press the veneer and note the bleed-through after it's out of the press. If the spread thickness is too heavy or too scant, you can adjust it accordingly for subsequent pressings. Keep a record of effective spread thicknesses for future reference.

If you don't want to use MSGL specifications to control spread thickness, you can use a film thickness gauge to check the thickness of a glue film after you apply it. You can buy a calibrated film thickness gauge or make your own. I make simple, uncalibrated gauges like the one in the photo below left and use them to estimate and compare film thicknesses from spread to spread rather than closely regulate them. This is not as precise as using a calibrated gauge or the MSGL system, but it's more precise than simply judging film thicknesses by eye.

Penetration   The glue you apply to a joint must thoroughly wet and penetrate its gluing surfaces in order to produce a good bond. The goal of wetting a gluing surface is to saturate it and to displace any air in the wood pores. Air that isn't displaced by glue becomes entrapped in the bond layer and can weaken it. Because of their chemistry or viscosity, some glues wet and penetrate better than others. You may need to switch glues when bonding workpieces that have difficult-to-wet surfaces.

Wood surfaces are hard to wet either because they have a dense pore structure (e.g., maple) or because they are hydrophobic, which means that they have a tendency to reject water (e.g., oily tropical woods). That's why non-water-based glues such as epoxy are preferred for gluing woods like teak or apitong.

To check the thickness of a wet adhesive spread, you can use a thickness gauge. This simple, shop-made gauge is cut from a milk jug.

CLAMPING

Glue joints aren't clamped simply so that your work will look tight and clean. They are also clamped to facilitate glue bonding. Most glues develop good cohesion only when pressed into a thin, even layer; they develop good adhesion only when the gluing surfaces of the work are in close, intimate contact with the bond layer.

Excessive clamping pressure should not be used to make up for joints that fit poorly or are improperly aligned. Instead, the joints should be refined or repaired as needed so they can be glued up properly without excessive pressure. When you overclamp a poorly fit or misaligned joint, you add unwanted stress to the assembly both during the pressure period and after the clamps are released, as the work tries to return to its previous state. You can also torque assemblies out of proper alignment and starve glue joints by applying too much clamping pressure. Use only as much pressure as is needed to close a joint properly. For most well-made joints, light to moderate pressure should suffice.

Cauls Clamping pads, blocks, and cauls are important for three reasons: They keep clamps from damaging the work; they allow you to put clamps on unusually shaped, hard-to-clamp workpieces; and they help direct and distribute clamping pressure. Size your cauls properly to the work at hand. If they're too big or to small, they can cause clamped work to become misaligned or racked out of square. In general, make cauls as big as the work will allow so they will distribute clamping pressure more evenly and thoroughly to the work and allow you to use fewer clamps.

Cauls should be simple, but take the time to make them effective and easy to use. You can rabbet cauls or add ledger strips so they hook around work and stay in place during clamping. You can tape or spot-glue cauls in place on the work with hot-melt glue to free your hands for clamp placement, or you can use hot melt to glue temporary tabs or ears onto cauls so they can be held in place by secondary clamps.

Pressure To glue your work accurately, you have to direct clamping pressure carefully. Orient the pressure to the width and thickness centerlines of the glued parts that are captured between the clamp heads, such as frame rails or drawer fronts. Check your work for

You can use a sprung (concave) caul to glue solid edging to a plywood shelf. With sprung cauls, glue-ups require fewer clamps (in this case only one).

Glued work can be heated simply and cheaply with readily available materials and equipment, such as the sheet plastic and electric heater shown here. Always use extreme care and caution because of potential fire hazards.

and clamping stage and a setting and curing stage. I often do whatever I can to extend the amount of time I have to apply glue and clamp the work, and then do anything I can to shorten the amount of time the work has to remain clamped. When using hide glue, for instance, I prewarm the work before gluing it, then place it in a cool area once it's clamped so the glue will gel more quickly. With adhesives such as PVA, urea resin, and epoxy, the tactics are reversed. I apply those glues in cool temperatures, then harden them quickly by heating the work with lamps and space heaters while it's in clamps.

Adding heat not only shortens the clamp times of many adhesives, but can also improve their penetration and ultimate properties. However, significant improvements in either area require substantial amounts of heat—well over 100°F. Most small shops can't (or shouldn't try to) produce this kind of heat. For my purposes, I'm happy with the 80°F to 90°F I get from my low-tech heating methods. If you decide to heat glued work, do so with extreme caution because of the potential safety hazards. Your other major concern should be to avoid harming the glued work itself by subjecting it to prolonged periods of heat and dryness, which could cause degrade or dimensional change in your materials.

For production and commercial operations, specialized glue-curing devices such as hot presses and high-frequency units are available, and they greatly reduce clamp times. Even the smallest of these devices can cost thousands of dollars, though, which makes them too expensive for most small shops. If you're familiar with heat sources and controls, you can make your

accuracy as you glue up. If you're edge-gluing, check with a straightedge and by sighting along the ends of the boards for flush joint alignment and flatness across the width of the work. Use a straightedge, a square or bar gauge, and winding sticks to make sure that frames and leg-and-rail assemblies are flat and square, and use a bar gauge to square up carcase assemblies.

If your joints are well made, directing the pressure along centerlines should bring your work together with proper alignment and orientation. If it doesn't, shift the pressure slightly off the centerlines by moving either the clamp, the caul, or both. If this doesn't square an assembly, clamp diagonally across the work to bring it into square.

Clamp time If you want to control clamp times for different gluing jobs, the best way to do so is through the strategic use of your shop and equipment. A glue-up has two distinct stages: an application own low-cost glue-curing devices from individual components (see the photo at right), or you can have an applied heating specialist help you design and make a system.

CLEAN-UP

In theory, if you have applied the right amount of glue and clamped your work properly, you shouldn't have much squeeze-out to contend with. In real life, gluing isn't always that orderly. You have to be prepared to clean up any amount of glue at any time.

It's important to develop systematic, effective clean-up methods for three reasons. First, you'll be able to produce work with surfaces that are free of excess glue. Second, you'll be able to control the amount of time you spend cleaning up. Finally, you'll be able to minimize the hazards associated with excess glue and its removal.

In most cases, cleaning up your work should be a primary concern, as any excess glue left on the work can ruin its appearance once stains and finishes are applied. Remove excess glue from your work by the easiest and most thorough means possible. Water-based glues such as hide glue and PVA are easier to clean up when they are still wet or slightly gelled. Non-water-based glues such as epoxy and polyurethane glue are easier to clean up after they have hardened.

How and when you remove excess glue from your work are often dictated by the situation. For instance, PVA squeeze-out can be left to harden on an edge-glued tabletop if you're going to plane the top after gluing it up. But PVA squeeze-out should be cleaned from the inner surfaces of carcase and

Gluing curved, tapered staves into a vessel form with a shop-made press. The electrically heated aluminum shoe reduces the clamp time to minutes. (Photo by Richard Scott Newman.)

frame assemblies while it's still wet because it is difficult to remove hardened glue from those areas neatly and thoroughly.

In some cases, you can prefinish workpieces before gluing them up if you wish. Doing so usually makes clean-up easier because most glues don't penetrate or adhere well to finished surfaces. Alternatively, you can tape off the areas around glue joints as shown in the photo on p. 81 to isolate them from squeeze-out. With either method, you can remove the squeeze-out when wet or let it harden, as you see fit.

Remove any glue that may have gotten on your skin while cleaning up your work or while cleaning your gluing equipment. As you clean glue applicators, rolling pans, and other

To dispose of old glue, you can use a plastic "cat box" filled with sawdust. The glue and sawdust will harden into a cake that can be popped free and discarded.

items, try to generate as little glue-fouled waste water or solvent as possible. If you have any leftover glue to dispose of, do so in a safe, expedient, and legal manner. Don't discard any unused or old, unusable adhesives in an unmixed or fluid state because they might leach into the soil at a landfill. Instead, bring them to a local hazardous-waste collection site if there's one in your area, or safely convert them to cured inert solids in your shop before discarding them.

Water-based glues such as PVA can be converted to a solid by pouring them into a pile of sawdust to harden (see the photo above). Non-water-based adhesives such as epoxy can be mixed and poured into sawdust as well, or poured onto a sheet of polyethylene plastic and peeled free and discarded once they harden. Work in a well-ventilated area—preferably outdoors—when you convert adhesives for disposal.

Once they solidify, dispose of them in accordance with all federal, state, and local regulations that apply in your area.

Conditioning

Properly conditioning your glued work after it has been unclamped is an important part of successful gluing. For many woodworkers, there's a great urge to continue working on glued assemblies as soon as the clamps come off, but this should be avoided whenever possible. Instead, glued work should be left alone for a period of time so that the glue can reach final cure and so that any moisture added to the work during gluing can dissipate.

Conditioning doesn't have to be elaborate. Edge-glued and veneered panels can be stacked and stickered on a flat surface and weighted down if needed. Joined assemblies such as chairs and table bases should also be placed on a flat surface so they won't distort out of proper alignment. Bent laminations can be lightly clamped to a temporary holding form or left on the original bending form. You can warm your work beyond room temperature during conditioning if you want (unless it's glued with hide glue or hot melt) to make the glue cure more quickly and shorten the conditioning period. If you do warm glued work, allow it to cool to room temperature before working on it further. If you don't warm your work, try to keep your shop at 65°F or above while the work is being conditioned, and don't let the temperature vary greatly.

The amount of time required for proper conditioning depends on the job, the glue you used, and the conditions

MOISTURE-RELATED CONDITIONING PROBLEMS

Puckered joint

This stile and rail were joined with a floating tenon, using water-based glue. The joint faces, planed smooth while the wood still contained added moisture, puckered as the moisture dissipated and the wood shrank, creating visible surface depressions.

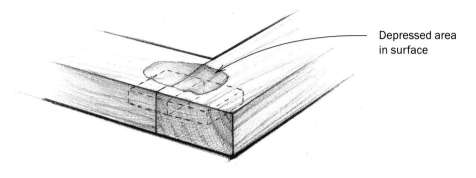

Depressed area
in surface

Sunken joint

The edges of this face lamination were planed flush soon after being glued with a water-based glue. As the wood released the moisture added by the glue, it shrank at the edges, causing a trough-like depression to form along the glue line.

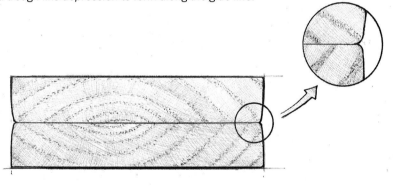

under which the work was glued. The simplest approach is to leave your work alone for as long as possible. Two to three days is an acceptable minimum conditioning period for most work. Any time you can allow beyond that will only benefit your work more.

If you don't condition your glued work, it won't fall apart, but you may encounter problems as you continue working on it or after it's completed.

Glued parts that are subjected to stress before the glue hardens completely may shift or change shape. Work that is machined or handworked while still containing moisture added during gluing may deform as that moisture eventually dissipates. Moisture-related conditioning problems can take many forms, such as puckered and sunken joints (see the drawing above).

11

Edge and Face Gluing

In many woodworking projects, the first glue-ups are usually edge joints and face laminations, which build workpieces up to size for other procedures, such as dimensioning, shaping, and joinery. Solid planks might be edge-glued to make door panels and carcase parts, face-glued to make turning blanks, or resawn into thin strips and bent into laminated curves. Workpieces cut from sheet stock are also routinely edge- and face-glued, and thin sheet goods are often used in bent laminations in place of solid wood or veneer.

Always make the best edge and face glue joints you can because they are often the cornerstone of a project's structural soundness. The construction steps that come after these glue-ups depend on their integrity. They can also be the hardest glue joints to fix if they fail. Repairing a broken mortise-and-tenon joint is easy compared to regluing a separated face lamination in a leg or a

popped edge joint in a tabletop. Good edge and face joints depend on workpieces that are properly dimensioned and have well-prepared gluing surfaces.

Solid Stock

Our woodworking ancestors didn't have to do much edge or face gluing. In fact, with some types of furniture, they didn't have to do any. That's because they had ready access to an abundant supply of massive, top-quality lumber, like the 3-ft.-wide boards that Philadelphia Chippendale piecrust tea-table tops were cut from or the 5-in.- and 6-in.-square timbers that Victorian bedposts were turned from. As woodworkers today, we usually have to do lots of edge and face gluing because many of the boards at a typical lumberyard are narrow (8 in. and less), and good lumber is often hard to find in thicknesses greater than 2 in. Unfortunately, much of today's commercial lumber can be difficult to edge- and face-glue successfully because it is often dried as quickly as possible under drastic conditions, a process that adds stress to the wood and makes it more unstable.

EDGE GLUING

Keeping two solid boards glued edge to edge takes a fair amount of adhesive strength. Each board is a package of mechanical forces that are in constant flux, and these forces meet at the edge joint. It's a challenge for a cured bond layer to hold these forces in check because the surface area of an edge joint is small in relation to the dimension and mass of the glued boards.

If you take the time to dimension and surface your stock properly, your effort will be rewarded with simpler, easier glue-ups. Well-prepared solid lumber is a breeze to edge-glue. Joints close neatly, less clamping pressure is needed, and you won't have to rely on dowels, splines, or biscuits to align the edges of boards for gluing because boards that are flat and true fall readily into the same plane.

Avoid adding dowels, splines, or biscuits to edge joints whenever possible. Using them takes time and effort, makes processing the glued work more complicated, and makes future repairs more difficult. Also, avoid tongue-and-groove joints and router or shaper-cut joints. These profiled edge joints register mating boards nicely and increase the amount of glue surface, but they're completely unnecessary, and are almost impossible to repair. In most cases, the tightest, strongest edge joints are those that are the simplest, most

The glue joint in this shop-made feed roller has partly separated because the gluing faces weren't well prepared and because the moisture content of the wood was too high when it was glued up.

Keyed, or profiled, edge joints cut on a shaper make flush edge gluing easy, but they're tough to repair when they pop open.

straightforward, and most cleanly made. There are some cases, however, where you can (and should) improve an edge joint by adding a spline or other type of connector to it.

Here's a rule I follow with edge joints: The thicker and/or less stable the lumber being glued and the less rigid the glue being used, the more helpful it is to add connectors. For example, if you are

SPRING-JOINTED EDGES

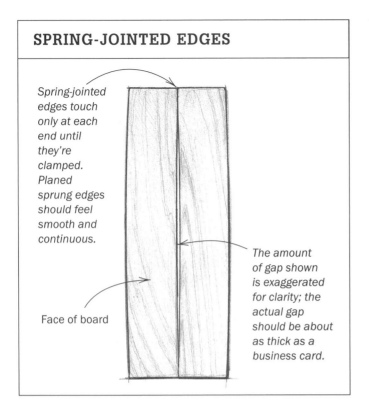

Spring-jointed edges touch only at each end until they're clamped. Planed sprung edges should feel smooth and continuous.

Face of board

The amount of gap shown is exaggerated for clarity; the actual gap should be about as thick as a business card.

edge-gluing a workbench top from 8/4 maple using consumer-grade white or yellow PVA glue, you should spline the edge joints. If you don't, the joints will probably creep because the PVA glue won't be able to resist the seasonal movement of the thick lumber by itself.

Jointing edges When I began woodworking, I had to rely on a jointer or a table saw to joint solid lumber. As I gained more experience, my edge joints improved dramatically because I learned how to use a hand plane. Hand-planing the gluing surfaces makes a tremendous difference in the quality and appearance of an edge joint. There's no need to become a total handwork fanatic, though. I use machinery to establish a gluing edge, then perfect it with a hand plane. I use the hand plane only when I need the best edge joint I can get. For decent ordinary edge joints, a well-tuned jointer is fine.

Spring jointing Planing the edge of a board accurately takes some practice. Once you have it down, you can add a bit of sorcery to your edge preparation by planing spring joints, which are slightly concave from end to end. Two boards with sprung edges touch only at the very ends as they're placed together (see the drawing above left). When clamping pressure is applied (see the photo at left), the joint closes evenly, with extra pressure added to the joint at each end. This pressure keeps the ends of the joint closed tightly, so that they will resist the common tendency to pop open over time due to seasonal dimensional changes in the wood.

To close a well-made spring joint whose gluing edges have been hand-planed slightly concave from end to end, all that's needed is one clamp placed at the midpoint of the joint.

FACE-GLUED SURFACES

Planer snipe (exaggerated for clarity)

Planer marks on gluing face

Edges of boards

Ends of boards

Hand planing the gluing faces of solid wood for face laminating removes mill marks and planer snipe, and greatly improves the quality of the bond.

END GLUING

Solid lumber can be glued end to end, but simple end joints are far weaker than simple edge joints. End-gluing solid stock so that it has an acceptable degree of strength is more involved than edge gluing. The traditional way to join the ends of boards is with a scarf joint. Scarfing is an effective and versatile joinery method because you can vary the design of scarf joints to suit the demands of your work. Another good method, which industry uses frequently, is to cut tapered finger joints in the ends of the boards with specialty router bits or shaper cutters.

Squared ends of boards that are simply butted together should have added mechanical connections that cross the joint lines, such as splines or biscuits. Both of these are quick and convenient to add, but neither is as strong as a scarf or a finger joint. To make any end joint as strong as possible, machine the ends of boards crisply and smoothly and avoid glazing or burning the end grain.

FACE GLUING

Face joints contain much more glue surface area relative to the dimension and mass of the glued workpieces than edge joints do. Nevertheless, face joints are still subject to plenty of stress from the constantly fluctuating glued lumber. The less dimensionally stable the lumber, the more stress it will impart to a face joint.

To prepare solid boards for face gluing, I machine both faces of each board flat and true, then hand-plane their gluing faces. Hand planing is well worth the effort because it greatly improves the gluing surfaces. It removes machine marks and irregularities such as planer snipe as well as any grease or oil that may have transferred from the machines to the wood. Hand planing also cuts through any surface compression, burnishing, or glazing caused by machining and leaves gluing surfaces with freshly opened pores that an adhesive can penetrate readily.

Well-prepared face joints don't require lots of clamps or pressure. This joint closed tightly with just the few lightweight clamps shown. For large face laminations, use heavier clamps and cauls or a veneer press.

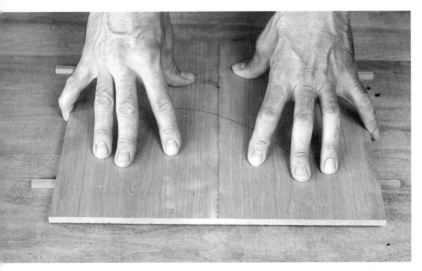

Traditional edge-gluing methods are simple and very effective. All you need are boards with neatly hand-planed edges and some hot hide glue. The joint is rubbed together and held fast until the glue gels, so there's no need for clamps. This is the easiest way to edge-glue thin boards.

ADHESIVES TO USE

Almost any permanent woodworking adhesive will hold a well-made edge or face joint together. The glue you use should be determined by how you want to work and by what type of work you're doing. If you're making heirloom furniture, you should consider using a reversible natural glue such as hot hide glue so the work will be easy for future generations to repair.

You can produce good hide-glued edge joints without clamps by using traditional methods. This approach is especially useful with thin stock, which is hard to clamp. Hand-plane gluing edges straight and true (no spring jointing) so that they mate well with hand pressure only, then apply hot hide glue to one or both of the edges and rub them together. The rubbing and hand pressure squeezes out the excess glue, while the glue's initial tack and rapid gelling hold the joint together. As the glue cures, it loses moisture and shrinks, which tightens the glue line.

The fastest and easiest way to bond edge and face joints is to clamp them up using PVA glue. Creep (see the sidebar on p. 106) is a concern when edge gluing and face laminating with PVA. Lower-grade PVAs cure to a relatively soft glue line, which can allow edge and face joints to creep, disturbing their flush alignment. When using PVA glue, the thicker the lumber that you're edge-gluing or the wider the lumber that you're face-gluing, the more likely creep is to occur. The easiest way to reduce or eliminate creep is to use a high-grade industrial PVA glue, which cures to a more rigid glue line.

Another good option is to make PVA glue more rigid by adding some mixed urea resin glue to it. Start by adding from 10% to 20% urea glue to your PVA and experiment to find the best mix proportions for your work. You can also add PVA to urea resin glues to make them less brittle and more durable, which enhances their usefulness for edge and face gluing. Start with 10% to 20% PVA, and experiment.

Polyurethane glue is also a good overall choice because it contains no moisture, is easy to clean up after it hardens, and has excellent cured properties when used in well-fit, tightly clamped joints. Epoxy is not quite as good an overall choice despite its superior cured properties because it's harder to clean up and works best in visibly thick bond layers, which are not desirable in face and edge joints.

For the best results when end-gluing solid stock, choose glue based on how the ends of the boards are joined. A good scarf joint can be successfully glued with any of the glues just discussed. Machined finger joints are best glued with a PVA or polyurethane glue. For splined end joints, use a PVA if the splines are nice and tight, or use epoxy if the splines are a bit loose and gaps need to be filled. For end joints connected with biscuits, a PVA glue works best. With any end joint, prewet the highly absorbent end grain of each board with glue before applying a full layer of glue so that the wood doesn't wick too much glue out of the clamped bond layer and starve the joint.

Bent Laminations

Bent laminations are the ultimate test of glue in a woodworking shop. They are basically a battle of forces, pitting chemistry against physics. Strips of wood that are laminated into a curved shape will attempt to flex back to their former flat state when the bend is released from the form; this generates a sizable mechanical force. The only thing countering this force is the ultimate strength and rigidity of the cured glue in the lamination.

Detail of chair by John Henry Belter, c. 1850. The entire chair back is a hide-glued rosewood bent lamination that is pierced and carved. Lamination lines are visible along the edge.

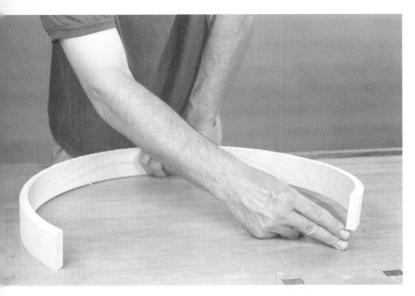

This bend was laminated from many strips of 1/16-in. ash veneer using PVA glue. The bend exhibited little springback after being unclamped, but twisted out of plane as it released all the moisture it had absorbed from the glue.

Whether it's made of sheet stock, solid wood, or veneer, the success of a bent lamination depends on the relationship between the thickness of the laminae and the amount and type of glue used during bending. If thinner laminae are used, the bend will exhibit less springback when it's released from the bending form. On the other hand, it will also have a higher glue content because more layers will be needed to create the desired thickness of the bend. This may make the bend more subject to planar and dimensional change as the glue cures.

Successful bent laminations strike a balance between the need to minimize springback and the efficiency of using as few strips as possible to make up the thickness of a bend. Carefully prepared gluing faces are critical. Laminae that have heavy machine marks or smudges of dirt or oil will not yield decent, long-lasting bent laminations.

When you force wood to hold a glued curve against its will, you've got to use an adhesive that's up to the job. There are four main properties to consider as you choose glue for bent laminating: strength, rigidity, moisture content, and closed assembly time. Ideally, you want high strength and rigidity without brittleness, a low moisture content, and ample closed assembly time so that you can manipulate and clamp the work comfortably after the laminae are coated with glue and stacked. Moisture content is especially important because the moisture in water-based glues can take a long time to evacuate from a laminated bend, and the bend can warp or shrink as the moisture dissipates. Tight bends can actually "toe in," or contract to a tighter radius, as they lose moisture.

Hide glue may seem like an excellent adhesive for laminating bends because it's strong and rigid. However, it has a high moisture content, and it gels so quickly that applying it fast enough to be able to clamp up a bend would be impossible under normal circumstances. Few woodworkers today would even think of using hide glue in a bent lamination, even though the gel rate of hide glue can be retarded (see p. 29).

PVA is a popular choice for bent laminations because it's convenient, ready to use, and very familiar to most woodworkers. You can use PVA for bent laminations, but the results you get will be inconsistent. Lower-grade PVA glues have little rigidity, and bends that are laminated with these glues won't be able to hold their shape after they're released from the form. They will spring back

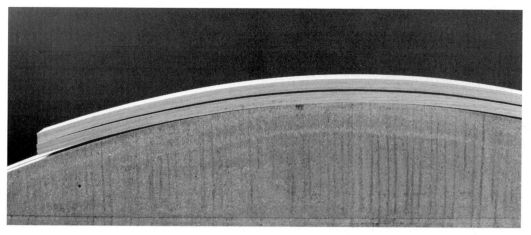

Here are two similar cured bent laminations resting on a bending form. The bottom piece was bent with PVA glue; the top piece was bent with two-part urea resin glue. The urea bend has less springback and contains less residual moisture.

significantly and may twist as they release the moisture they absorbed from the glue. If you laminate bends with PVA, use an industrial high-solids formula, which will have a lower moisture content and dry to a harder, more rigid bond layer.

Urea and resorcinol resin glues are the best adhesives for laminated bends. They're extremely strong and rigid, they're easy to apply with rollers or brushes, and they usually allow ample time for spreading, assembly, and clamping. Of these glues, two-part urea resin glue is the best choice because it contains less moisture than one-part urea resin glue and most resorcinols, and it cures to a neutral tan color rather than a conspicuous dark red line, as resorcinol does. Two-part ureas are also generally less brittle and likely to craze than one-part ureas.

Polyurethane is also a good choice for bent laminations, and is a better all-around choice than epoxy for the same reasons discussed on p. 121. Polyurethane glue allows ample assembly times, its foam-out is easy to machine away after a lamination cures, and it doesn't dull cutting edges as quickly as urea and resorcinol squeeze-out does. Epoxy is fine for a gentle bend where only light clamping pressure is needed, but it is not an ideal choice for a tight bend where the great amount of pressure required is likely to drive too much epoxy out of the glue lines, starving the lamination.

When using any of the above adhesives (other than hide glue) for bent lamination, you may need or want to heat the work (see the photo on p. 112). Adding heat will help the glue cure more quickly and develop superior ultimate properties. Use extreme care and caution whenever heating any work, and never leave heat sources unattended while they're in use.

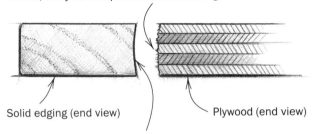

SOLID EDGING OF SHEET GOODS

Raised, fuzzy core of panel after machining

Solid edging (end view)

Plywood (end view)

Hollow of solid-wood gluing surface made with
a radiused plane iron (exaggerated for clarity)

*The cores of MDF, plywood, and flakeboard are often made
fuzzy and raised by machining. Hand-planing the gluing
surface of the solid edging improves the joint. For a tight
glue line, plane a slight hollow into the edging if you want.*

Sheet Goods

Most edge gluing of sheet goods involves
the application of solid-wood or
manufactured lippings to the edges of
panels cut from sheet stock, such as
cabinet sides or tabletops.

To get the tightest possible edge joint
between sheet stock and solid-wood
lippings, machine the edge of the sheet
stock as carefully and crisply as you can.
Then hand-plane the gluing surface
of the solid edging. Hand planing the
edging helps make up for any
deficiencies in the machined edge
of the sheet stock. This is much better
than using extra clamping pressure to
force less carefully prepared gluing
surfaces together.

Solid wood that's glued to the edge of
a man-made panel is more likely to
change dimension with seasonal
variations in relative humidity than the

panel is, which may cause the glue joint
to creep over time. Again, the best way
to combat creep is to use a rigid glue.
You can also use splines or biscuits if
you want, but I don't bother using them
when gluing simple solid-wood edging
that will be leveled flush to the panel
surfaces afterwards. I just make the
edging oversize and clamp it to the panel
edge so that it stands proud of both
panel faces. I remove the squeeze-out
after it dries while trimming the edging
flush to the panel surfaces with a router.
I always use biscuits whenever I edge-
glue two manufactured panels together,
though. The biscuits help keep the
panels flush as I clamp them together so
I don't have to level them flush after
gluing them up, which is tedious and
difficult to do well.

FACE-LAMINATING SHEET STOCK
Sheet goods such as MDF and
flakeboard are manufactured with true,
flat faces that can be easily glued to each
other. Plywood, on the other hand, is a
product whose overall quality is steadily
declining. These days, plywood sheets
are rarely flat, and face veneers often
have surface variations that can
make good face gluing difficult. When
a plywood face lamination is critical,
I lightly skim the mating face veneers
with a hand plane to knock down any
high spots before the glue-up.

ADHESIVES TO USE
The edges and faces of plywood,
flakeboard, MDF, and other sheet goods
are very porous and can absorb a lot of
glue. Don't use low-viscosity glues
because they'll soak in too much and
leave joints starved. Apply a heavier
spread thickness of glue than you would

TRIMMING SOLID EDGING ON SHEET STOCK

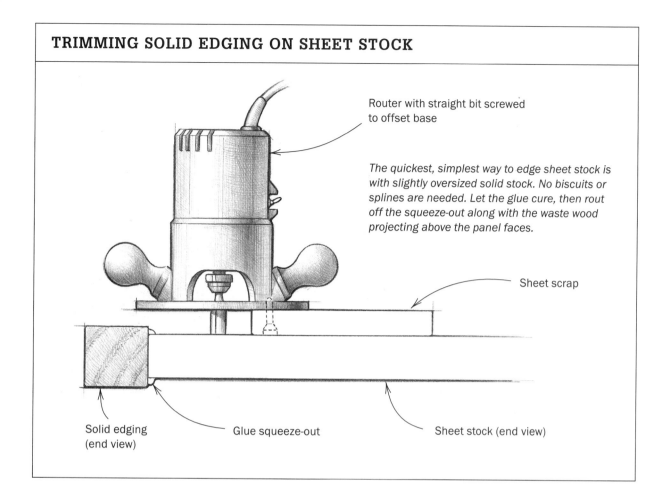

Router with straight bit screwed to offset base

The quickest, simplest way to edge sheet stock is with slightly oversized solid stock. No biscuits or splines are needed. Let the glue cure, then rout off the squeeze-out along with the waste wood projecting above the panel faces.

Sheet scrap

Solid edging (end view)

Glue squeeze-out

Sheet stock (end view)

for edge and face joints in solid wood. Some of the glue will soak into the panel core while the joint is open, leaving a suitable amount of glue on the surface for a good bond layer when the joint is closed.

An industrial-grade slow-acting PVA glue is an excellent choice for panel-to-panel edge joints because it allows long assembly times for careful edge alignment, swells biscuits properly, and cures to a fairly rigid glue line. It will also allow you enough time to clean up the squeeze-out while the glue is still wet. If you do clean wet PVA from

edge-glued panels, use as little water as possible, because excess moisture can swell and degrade MDF and flakeboard.

For applying solid-wood edging to sheet stock, slow-acting PVA is again a good choice if you're using biscuits to align the work. If you're gluing the edging without biscuits, use a regular-speed PVA so you can get edged panel parts in and out of clamps quickly.

Most other glues are secondary choices for edge-gluing sheet goods because they require mixing, take too long to cure, or are too expensive, toxic, or difficult to apply and clean up.

The face veneer of plywood can be planed for face laminating if it's thick enough to allow such work. Smoothing out surface variations improves the quality of the glue joint.

PVA is also good for face-laminating sheet stock. There are some reasonable alternatives as well, including casein glue, urea resin glue, polyurethane glue, and a mixture of PVA and urea resin glue. All of these adhesives allow a longer open time, cure to a rigid glue line and aren't too expensive, although some brands of polyurethane are still overpriced. Some polyurethanes are also highly viscous, which makes them hard to spread over large surfaces. Polyurethane is the only glue in the group that's moisture free, which makes it worth considering for big panel stock face laminations, where any added moisture could cause warpage after the work was unclamped.

Conditioning Edge and Face Joints

It's important to condition edge and face joints by letting glued work remain at rest for a period of time after it is unclamped. The work should be conditioned if moisture and/or heat have been added to the stock during a glue-up. Conditioning allows the glue to cure and lets the added heat or moisture dissipate from the work so it can reach a point of equilibrium.

To condition edge-glued work, you can lean it vertically against a wall after unclamping it or sticker and stack it on a flat surface. Either way, keep it out of direct sunlight. Big solid or sheet-stock face laminations can be stacked on a flat surface and weighted down with cinder blocks or sandbags. Bent laminations don't need to be confined during conditioning, unless you think the bends may twist or deform. In that case, you can clamp them to a temporary holding form or the original bending form to help keep them in shape. Reclamping a bend won't eliminate any springback that has already occurred, but it may prevent more springback from occurring.

12

Veneering

As popular as veneering has become over the last 20 years, to many woodworkers, it still appears too daunting to tackle. While veneering does require careful preparation, good organization, and bucketfuls of glue, it's not hard to do, and it's far too useful and important a part of woodworking to neglect.

In this chapter, I'll discuss various kinds of substrates that are used in veneering, how to prepare substrates and veneers for gluing, and what glues are best to use for different veneering situations. The chapter also contains guidelines for successful veneering glue-ups and a discussion of various effective ways of veneering without clamps and presses.

Choosing Substrates

Before you begin veneering or overlaying projects, it's important to choose substrates that are well suited to the work. You can use solid wood or manufactured sheet goods such as plywood, or you can fabricate your own custom substrates. Whichever you choose, your main objective is to provide the work with a foundation that is durable and dimensionally stable. The first thing to determine is whether the substrate will have a hollow core or a solid core. Solid-core substrates are the standard choice for veneering and overlaying, but hollow-core substrates such as honeycomb sheet stock are becoming increasingly popular because they allow you to construct veneered or overlaid panels that are strong, rigid, and lightweight, as well as stable.

Veneering is an important part of woodworking. It can be done well without complicated setups and expensive equipment. Here, a curved table apron is being veneered using handscrews and flexible cauls. (Photo by Allan Breed.)

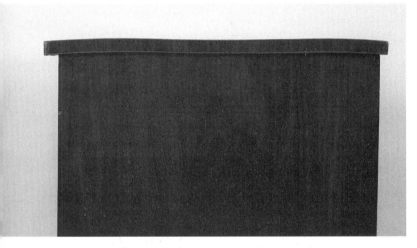

The top of this chest is made from solid wood, which is not an ideal substrate for large veneered panels. It has cupped downward because it was veneered on one side only. Both sides should have been veneered.

Solid wood has been used as a veneering substrate for centuries, and it's still in use today. It's not an ideal substrate because it doesn't have good dimensional stability. It shrinks and swells across its width on a seasonal basis and can cup, bow, and twist. Nevertheless, solid wood can be an effective substrate if it's properly chosen and used for moderately sized work. Carefully dried, clear-grained woods such as pine and basswood are good substrate choices because they're soft, easy to work, and glue well. However, solid-wood substrates that are thicker than ¾ in. to 1 in. or wider than 8 in. to 10 in. are likely to distort and cause problems after being veneered. I wouldn't use a solid-wood substrate for anything bigger than a dresser-drawer front.

If you use sheet goods for solid-core substrates, buy the best grades that you can. Cheap plywood is neither flat nor uniform, and inexpensive flakeboard, such as floor underlayment, is coarse and crumbles easily. Furniture-grade flakeboard, MDF, and lumber-core plywood make the best sheet-goods substrates.

The best prefabricated hollow-core material to use for woodworking is resin-impregnated paper honeycomb core stock. It's easy to cut and bonds well to many different face materials with a variety of adhesives. Although you can buy honeycomb core stock with faces that are preskinned with paper, I prefer using unfaced, open-cell honeycomb, so I can glue on face skins of my own choosing, which can be as thin as plastic laminate or ⅛-in. plywood.

If manufactured substrate materials won't meet the demands of your work, you can make your own substrates. There are four good practical ways to make substrates in a small shop. First, substrates can be made from face-laminated sheet materials or solid wood that has been resawn into thin plies. This method is often used to create curved substrates, and it is also a good way to build up custom substrate thicknesses.

Second, you can make your own lumber-core plywood substrate stock by edge-gluing narrow strips of solid wood into a core whose length, width, and thickness you determine as needed. Both faces of the core are then crossbanded with a layer of veneer whose grain runs at 90° to the core grain. This restricts the lateral seasonal movement of the lumber core and makes the substrate much more dimensionally stable than simple solid-wood substrates. Face veneers, in turn, are glued perpendicularly to the crossbanding so that their grain runs parallel to that of the lumber core.

Third, you can make curved substrates out of solid wood by using bricklaid construction, which is a very useful traditional technique. Bricklaid curves are built up of short, narrow lengths of wood that are butt joined end to end along an arc and stacked in edge-glued courses with their end joints staggered, just like bricks in a wall. After the glue dries, the faces of bricklaid cores are bandsawn and faired into smooth curves, then crossbanded with veneer, just like lumber-core plywood.

Finally, you can make your own hollow-core substrates using torsion-box construction, a simple but highly effective system that offers lots of design possibilities. The core of a torsion box is a gridwork made from solid lumber or sheet-stock parts that function like a honeycomb core. The grid parts are butted to each other and stapled together to hold them in place until face skins can be positioned and glued on. When both face skins are bonded to the gridwork core, the assembled torsion box is much stronger and more rigid than either the core or face skins are by themselves.

Preparing Substrates

No matter what type of substrate you decide to use, you should make sure its surfaces are properly prepared before gluing to ensure the best possible results. There are two ways to prepare a substrate surface: scoring and sizing.

Veneer generally bonds better to substrates that have been scored or roughened before gluing. (Adhesives such as polyurethane glue and epoxy actually require a roughened surface in order to bond well.) You can roughen a substrate surface by scuffing it with sandpaper, or you can score it the traditional way, with a toothing plane (see the photo below), or with a hand-held toothing blade.

Scuffing or toothing a substrate surface does three things: It keeps the veneer from sliding around on the

Preparing a substrate with a toothing plane. Properly done, toothing is a surface treatment that can improve bonding in some situations.

substrate while it is being pressed down, it improves mechanical adhesion, and it allows more glue to be held in the bond layer while still allowing intimate contact between the substrate surface and the pressed veneer. Using a toothing plane is the most precise way to prepare a substrate for gluing. Making one full, even pass over a substrate will leave its surface flat and partially intact when you're done. Sandpaper, on the other hand, abrades the entire surface, and can alter its flatness if you're careless. Toothing is best used when you're using hide glue or epoxy, or when you're veneering onto solid wood or plywood. Scuffing is best used when you're using PVA, urea resin glue or polyurethane glue, or when you're veneering onto flakeboard or MDF.

Substrate surfaces that are highly porous and absorbent, such as the end grain of solid wood, should be sized with a preliminary application of glue. Sizing fills pores and partially seals these surfaces so they won't absorb too much glue and starve the bond layer. There are two ways to size a substrate: You can apply the size as a prewet coat just before applying the full, regular coat of glue for the veneer, or you can apply the size in advance and let it dry. With either method, thin the sizing glue, if possible, so it will penetrate the substrate better. If you plan to let the size dry, wipe off any excess while it's still wet, and scuff-sand the substrate after the size dries so the full, regular coat of glue will adhere well to the sized surface.

Preparing Veneer

Veneer needs to be prepared for gluing just as substrates do. Begin preparing sliced veneer by testing it for wetability, which is done by flicking a few droplets of water on its surface. If the water beads up instead of soaking in, it means that the veneer surface was compressed and burnished during the slicing process, which will prevent glue from wetting and penetrating the surface properly. Poor penetration will cause the veneer to delaminate after being bonded to a substrate. Compressed surfaces should be scuffed with sandpaper to open them up so glue can penetrate them. Scuffing is actually a good general veneer preparation method, because all veneer surfaces oxidize and collect grime as they age, which makes them more difficult to bond properly.

To prepare thick, commercially sawn veneer, or veneer that you have resawn from solid lumber in your shop, remove

Before gluing down sawn veneer, skim off the saw marks with a block plane. After planing, scuff the surface lightly with sandpaper to promote good bonding when the veneer is glued down.

most or all of the saw marks with a hand plane to level and smooth the gluing surface. You can then scuff the smoothed surface with sandpaper as you would for sliced veneer.

Veneers that are wavy, buckled, or fragile should be treated before being glued down. This is done by wetting the veneer with water, a conditioning solution, or a flattening solution, then pressing it flat to dry between two pieces of plywood, using layers of kraft paper or plain newsprint to absorb the moisture. Conditioner and flattener also act as sizes and help prevent bleed-through when the veneer is glued down. They will not hamper the veneer's ability to take a finish if used properly.

Water is fine for treating fairly well-behaved veneer. If your veneer is in rough shape, you'll need to use a conditioner or flattener instead. Veneer conditioner consists of alcohol, water, and glycerine. You can make your own, or you can buy ready-mixed conditioner from mail-order suppliers. Conditioner moistens and softens veneer to make it more pliant and workable, and less likely to degrade once it is glued down. Neither water nor conditioner will keep the veneer flat once it has been removed from the plywood flattening press.

Flattener does keep veneer flat once it's dry because it contains glue in addition to alcohol, water, and glycerine. Flattener can be hard to find at supply houses, but it's easy to make your own solution as follows, using volume measurements:

4 parts water
2 parts PVA glue (white or yellow)
1 part denatured alcohol
1 part glycerine

If you prefer to use a more traditional flattener, here's a recipe that includes hide glue:

5 parts water
2 parts hot hide glue (already mixed and heated)
1 part glycerine

Because flattener contains glue, you must use pieces of fiberglass screen to isolate the veneer so it won't stick to the paper and the plywood as you flatten and dry it. Whether you use water, conditioner, or flattener, change the newspaper often—several times the first day, then once a day for at least two more days. You must dry the veneer completely before gluing it or it will shrink and crack after it's glued down. Don't rush the drying or spare any effort in tending to the process. Once the veneer is dry, keep it stored flat in a top-weighted stack until you use it.

Wavy, buckled veneer should be treated before it is glued down. After being saturated with a shop-mixed flattening solution, this sheet will be pressed flat to dry.

Gluing Substrates

When you make up your own substrates, you should choose the right glue for each job because there are several types of substrates you can make, and the demands that are placed on substrates can vary from project to project. You may well find that in some cases, the adhesive you use to make a substrate is different from the adhesive you veneer it with.

FACE-LAMINATED SHEET SUBSTRATES

Since sheet goods are generally stable materials, the glue you use to face-laminate them into substrates isn't of critical importance. Although face laminating is a good way to use up aging or excess glue of almost any sort, I usually use PVA glue because it's fairly inexpensive, easy to apply with a roller, and sets quickly.

More important than your choice of glue is the composition of the lamination package. It should have a central core ply with a balanced (symmetrical) arrangement of outer plies. For example, if you need a 1-in.-thick substrate, glue a piece of ¼-in. sheet stock to each face of a ½-in. piece, instead of gluing a piece of ¼-in. material to one face of a ¾-in. piece. Unbalanced laminations usually bow or cup after being glued up.

SHOP-MADE LUMBER-CORE-PLYWOOD SUBSTRATES

When edge-gluing narrow strips into a lumber core for shop-made plywood, convenience is an important factor. PVA and polyurethane glue are the best choices because they're ready to use and easy to clean up after they dry. PVA allows you to build up core widths rapidly because it sets quickly. Polyurethane glue allows long assembly times and doesn't contain any moisture.

The best adhesives for crossbanding a lumber core are two-part urea resin glue and polyurethane glue, because they introduce little and no moisture to the work respectively, and they offer long assembly times so both faces of the core can be crossbanded at the same time. Both glues are also rigid once they cure, which is important, because crossband layers are meant to restrict the lateral seasonal movement of the solid wood in the core.

BRICKLAID CURVED SUBSTRATES

Hot hide glue is the adhesive of choice if you're gluing up a bricklaid curved substrate. Its high initial tack allows you simply to rub parts into place instead of having to clamp or nail them. With rub joints, you can assemble a bricklaid curve in a very short time. Hide glue is very rigid and creep resistant after it cures, which keeps the glue lines in a core from telegraphing through to the face veneer. A good second choice would be a high-speed-assembly PVA, which would allow you to build up a bricklaid curve almost as quickly as hide glue.

Hide glue is also a good choice for crossbanding a bricklaid core with veneer because it allows you to apply the crossband layer with a veneer hammer instead of clamps (see pp. 138-140 for a discussion of hammer veneering). If you don't want to hammer the crossbanding on, you can retard hide glue with urea as discussed on p. 29 to extend its assembly times, so you can glue the crossbanding down with clamps and cauls.

Equipment for Pressing Veneer

Three basic types of veneer presses are used in small shops—improvised devices, screw-type presses, and vacuum presses. Each of them has advantages and disadvantages. Improvised presses can be made from almost anything—the crude gravity press I started out with was made from sheets of flakeboard and stacks of water-filled five-gallon pails. Rigs like this are inexpensive and they work, but they can be bothersome to set up and take down, and they don't always distribute pressure evenly over a large gluing surface.

With a screw-type press, you can distribute pressure to the work evenly, provided that you arrange an ample system of cauls inside the press. Loading a screw press can take many precious minutes, and fast-acting glues may start gelling before you can begin applying pressure. On the other hand, you can cycle work in and out of a screw press fairly rapidly because the surrounding shop atmosphere circulates freely around the work, which helps most glues set and cure more quickly.

Vacuum presses deliver pressure that is distributed very evenly over the entire gluing surface. They don't require the same extensive system of cauls that screw presses do, so they're quicker and easier to load and activate. However, once glued work is sealed in a vacuum press, it's isolated from the shop atmosphere, and this affects how some glues perform. Adhesives that cure by chemical reaction, such as epoxy, work as well in a vacuum press as they do in the open air. But evaporative glues such as PVA can take longer to dry in a vacuum press because there's no surrounding atmosphere to help evacuate their moisture.

The absence of an atmosphere in a vacuum press can also affect the performance of moisture-curing adhesives such as polyurethane glue. Without any atmospheric water vapor available to help activate the glue, it has to rely solely on the moisture content of the glued materials to cure properly.

TORSION-BOX SUBSTRATES

Slow-speed PVA, urea resin glue, and polyurethane glue are all good choices for gluing the core and face skins of a torsion box together. They all have a long open assembly time, which you'll need to spread glue, close the assembly, and press it. PVA is the easiest to work with, but it isn't as rigid as the other two. It has the shortest required press time of the three, though.

Apply the glue to the core grid, not the face skins. The gluing edges of most core grids are at least ½ in. wide, and will hold enough glue to create a good bond with the skins. You can glue one skin at a time to the core, but glue down the second skin as soon after gluing the first one as you can.

Always use a top caul sheet when gluing down the top skin of a torsion box in a vacuum press. Because a vacuum press bag is powered by atmospheric pressure, it seeks out voids and conforms to shapes and contours. If you don't use a top caul, the vacuum bag will deflect a top skin that's thinner than ½ in. downward into the core grid's empty

When you skin one side of a torsion-box core in a vacuum press, the press bag conforms to all the voids in the core grid, providing even pressure overall.

spaces. The deflections become permanent visible depressions in the surface of the face skin as the glue hardens.

HONEYCOMB CORES

The two best glues for bonding face skins to honeycomb cores are two-part urea resin glue and PVA. When gluing face skins to honeycomb, apply the glue to the back side of the skins, not the core. If you use a veneer press to bond the skins to the core, don't use excessive pressure or you may deform the core cells. Be especially vigilant with a vacuum press. When vacuum pressing face skins onto honeycomb, use no more than 7 in. to 10 in. Hg. pressure and a use top caul sheet that overhangs the core by ¼ in. all around so the bag can't crush the perimeter cells.

Gluing Veneer

Although many woodworkers approach veneering by finding one method that they're comfortable with and then using that exclusively, there are many good ways to glue down veneer, both with and without the use of a press or clamps. It's a good idea to become familiar with more than one way to glue veneer, so you can use the most appropriate method for a given job.

ADHESIVES TO USE

In choosing which adhesives to use for veneering, it's worth considering the physical characteristics of veneer itself. First, veneer is hygroscopic, and it is affected by changes in moisture levels just like solid wood. It is also not uniform—it can vary in density, porosity, and surface condition, even within a single sheet. Finally, although veneer is generally pliant, some veneers can be quite fragile. Given these characteristics, veneer has to be bonded with a strong, rigid, durable glue line in order to remain well-attached to a substrate for a long period of time.

There are five types of permanent glues that are good choices for veneering: hide glue, PVA, urea resin glue, epoxy, and polyurethane glue.

Hot hide glue is strong and durable, but it tacks and gels quickly. Often it can't be used for work that must be clamped or pressed unless its gel rate is

Hints for Successful Veneering

As you get ready to glue down veneer, there are a few guidelines you should follow regardless of the veneering method you choose.

• Work carefully and safely because most synthetic glues can be hazardous and/or toxic, and veneering usually involves large surfaces and lots of glue.

• Balance veneered work by veneering both sides of the substrate. A substrate that's veneered on one side only will bow or cup. Veneer both sides at the same time whenever possible, especially when using a water-based glue. It's hard to keep a substrate flat when you add moisture first to one side, then to the other.

• When you need to mix glue, do so carefully and consistently, so each batch has the same viscosity and mix ratio. This will help ensure that all the glue lines in a multi-layer pressing will have the same thickness and cured properties.

• Apply glue only to the substrate, unless the procedure you're using calls for glue to be applied to the veneer as well. Apply glue in the proper spread thickness and as evenly as you can on each surface you veneer. You can control spread thickness by using a specialty glue spreader (see the photo below left), or you can apply glue in preweighed amounts or use a film thickness gauge (see the photo on p. 110). No matter how you apply glue, wet out the entire gluing surface. Don't depend on clamping pressure to help spread glue.

• Press the veneer to the substrate with controlled, even pressure, both as a further means of ensuring glue lines of consistent thickness and quality and as a means of controlling squeeze-out.

• Warm your work while it's being pressed, using the safest, simplest method possible, if your shop's temperature is below your glue's minimum temperature requirements or if you want to speed curing and reduce press times. A screw press can be tented with a plastic sheet and warmed with a space heater, and a vacuum press can be warmed with an electric blanket (see the photo at right).

Use heat sources with extreme caution, because they can be hazardous if misused. Warm the work to 80°F to 90°F only. If you heat pressed work beyond 110°F, it will warp, unless it is pressed with costly, sophisticated equipment.

• After veneering, check your work as soon as possible to make sure that every square inch of veneer is glued down properly. If you find areas that haven't adhered well, repair them as quickly as you can—with the same glue used originally. You can slit open blisters in the veneer, inject glue with a syringe, and clamp the area flat with waxed paper and a block of wood.

• Once the glue has hardened, clean up any bleed-through by scraping or sanding. Avoid using heat, water, or solvents as cleaners, because they may soften the bond layer or distort the panel.

A glue spreader will quickly coat large surfaces with the proper spread thickness for panel and veneer pressing.

An electric blanket is a simple, effective way to keep vacuum-pressed work warm so that synthetic glues will cure thoroughly and quickly.

Permanent vs. Non-Permanent Veneering Glues

The glues that are used for veneering can be divided into two basic categories—permanent and non-permanent. Most veneering adhesives are permanent. The non-permanent adhesives are hot melt, contact cement, and their variants, such as pressure-sensitive adhesive (PSA), a contact adhesive commonly known as "peel and stick."

Woodworkers choosing a glue for veneering are faced with a dilemma: Although non-permanent glues make veneering more convenient, permanent glues do a far superior job. Hot melt and contact cement simply can not bond veneer to a substrate with as much strength and endurance as an adhesive such as urea resin glue. Therefore, you have to choose between how easy you want veneering to be and how long you want your work to last.

This choice is a tough one for may woodworkers because convenience veneering has become a well-established system complete with its own specialty tools and materials, such as pre-flattened, paper-backed veneers. It is also widely discussed in publications and heavily promoted in catalogs. If you're not experienced with veneer, you could easily conclude that convenience veneering isn't just easier but is actually the best way to work, which is not the case.

Vacuum-pressing veneer glued with hide glue. Heat helps the glue flow well, but must be used cautiously to avoid damaging the vacuum bag. (Photo by Kim Schmahmann, courtesy Letitia Hafner; location: MIT hobby shop.)

retarded. To get longer assembly times, you can also switch to a lower-gram-strength glue and warm the substrate. Hide glue bleed-through is easy to scrape and sand after it hardens. Hide glue is also a superb pore filler that's very compatible with finishes. It's the only glue that can be used in hammer veneering (see pp. 138-140).

PVA has become such a popular veneering glue that glue companies now make PVAs especially for veneer work. You don't have to use a veneering PVA to get decent results, though. Cured PVA bleed-through scrapes and sands fairly easily. Higher-grade formulas are more sandable and more rigid when cured. Traces of PVA that are left on the veneer surface will hamper staining and finishing, but glue buried in the pores is

Gluing Overlays

An overlay is an applied face material other than veneer. Many different overlays, which are based on metal foils, plastics, paper, and other materials, are available. The most commonly used overlays in small shops are plastic laminates. They are much easier to glue down than veneers because they're uniform, rigid, durable, and impervious to moisture. Nevertheless, they have to be glued down carefully and skillfully in order to obtain good results.

The adhesives most often used to bond plastic laminates are contact cement, hot melt, PVA, and urea resin glue. In small shops, hot melt is typically restricted to small-scale applications, such as gluing plastic laminate to the edges of shelves. Urea resin glue is generally used only when an ultimate-quality bond is needed, which isn't that often. PVA and contact cement are much more frequently used with plastic laminates. PVA works well because most substrate surfaces, such as flakeboard, are porous (at least one part of a PVA-glued joint should be porous), and because the phenolic backing of plastic laminate is heavily scored to promote good bonding. Once it's cured, PVA is much more rigid and creep resistant than contact cement.

PVA does have some drawbacks, though. It's less heat resistant than contact cement (which makes it a poorer choice for kitchen countertops), it needs to be clamped or pressed to a thin bond layer, and it adds moisture to the substrate, which can cause it to cup or bow. For the best results, use a high-solids PVA formula with a low moisture content, and overlay both sides of the substrate to balance the assembly so it will remain flat and true.

If you use contact cement, you've got a choice between solvent-based and water-based cements. Solvent-based cement doesn't add any moisture to the substrate, but water-based cement does, and it also takes longer to dry before assembly, which will slow down your work rate. However, water-based cements contain a higher percentage of solids. In fact, some of the newer high-grade formulas have a solids content as high as 60%. They dry almost as quickly as solvent-based cements. This makes them a great choice, because they're neither flammable nor toxic, as solvent-based cements are.

When contact-cemented plastic laminates and substrates are assembled, the cement doesn't develop full strength unless pressure is applied to the work. Most people apply pressure with a hand-held rubber roller, but the amount of pressure you can generate with a hand roller is barely adequate. You can make the cement glue line much stronger by clamping or pressing the work, which squeezes the cement into a thinner layer. Just a brief period of pressure (even 30 seconds) will significantly strengthen the bond.

usually not a problem. You can tint the glue to make it color compatible with the veneer, if you wish.

Urea resin glues allow ample assembly times, have a high solids content, and cure to a very rigid glue line, which makes them ideal for veneering. They are more resistant to moisture and heat than PVAs and are easy to tint as needed. Two-part urea resin glue contains very little water and adds almost no moisture to the work. Urea resin bleed-through is hard, but scrapes and sands well.

Epoxy also allows long assembly times if mixed with a slow hardener. It has a very high solids content and a superior combination of strength and durability. It can be altered as needed with thickeners, tints, and other additives, and adds no moisture to the work. To veneer successfully with epoxy, scuff both the substrate and the veneer before gluing, and prewet porous substrate surfaces with epoxy before applying a full, regular coat of the same epoxy to the substrate. Then press the work, using just enough pressure to force the veneer flat against the substrate.

Polyurethane glue is convenient to use, although the thicker formulas are hard to spread over large surfaces. It allows long assembly times, contains no moisture, and cures to a rigid, durable glue line. Workpieces should be pressed with ample pressure to avoid foaming in the bond layer. The glue should be applied sparingly so it won't bleed through porous veneers too heavily. Polyurethane bleed-through is easy to scrape and sand, and glue traces in the pores won't hamper finishing.

VENEERING BY HAND

Using clamps and presses to glue down veneers produces great results, but can be time-consuming and tedious. Fortunately, there are several methods of gluing down veneer by hand that are faster and more fun than clamping and pressing, yet also yield durable results. It's worth mastering these methods to expand your technical repertoire and provide yourself with a wider range of options in any veneering situation.

Hammer veneering

Woodworkers have been hammer veneering for centuries and will be for centuries to come because it's a great way to glue down veneer. All it takes is some hot hide glue, a veneer hammer, and a bit of practice. In essence, the technique involves pressing the veneer onto the substrate with the hammer while using it like a squeegee to force excess glue out from between the veneer and the substrate. Hide glue tacks well

This European walnut cabinet by Eric Englander is hammer veneered from top to bottom, including the moldings. (Photo by Lance Patterson, courtesy Eric Englander.)

and gels rapidly as it cools, allowing you to position and firmly secure the veneer with deftness and precision as you manipulate the hammer. Hammering techniques may vary somewhat from woodworker to woodworker, but here are some things you can do from a gluing standpoint to ensure your own success:

• **Use the right glue at the right consistency.** A gram strength of 192 is a good starting choice. Glues with higher gram strengths gel more quickly and allow less open time. Glues with lower gram strengths gel more slowly, but also have a thinner consistency and tack less vigorously. Mix your glue in proportions recommended by your supplier to begin with, and adjust the viscosity as needed. The glue should be moderately thick and free-flowing, like syrup or gravy. Make it fresh for the job instead of using reheated glue.

• **Tooth solid-wood substrates that have dense surfaces with tight grain.** Toothing doesn't take long, and it keeps the veneer from sliding around as you hammer it. It also gives the glue somewhere to go as you hammer the veneer down; you can press the veneer close to the substrate surface even when the glue chills and thickens. Because hide glue shrinks as it cures, the veneer and substrate will be pulled tightly together.

• **Prepare wavy and buckled veneer with hide-glue flattener** (see p. 131). This makes it flat and pliant and presizes it with glue. When you cut veneer into pieces, keep excess and overhangs to a minimum. Veneer that hangs off the edge of the substrate will curl up and lift the veneer from the substrate. Warm the substrate (a clothes iron works best)

and your shop as much as possible before veneering to help extend the working time. Don't overheat the glue, and don't warm the veneer.

• **Coat both sides of the veneer and the substrate with glue.** Work quickly, but don't slather on the glue indiscriminately. Lightly wet the face side of the veneer with water after placing it on the substrate. Press the veneer onto the substrate with the hammer, working from the center outward in small sections. You should be able to hammer-veneer small workpieces before the glue gels. For bigger jobs, use a clothes iron to reheat the glue under the veneer if you need to, using the lowest heat setting possible. In general, keep reheating to a minimum. Each time you reheat hide glue, you weaken its structure and lower its cured properties.

• **Clean up the squeeze-out** from around the edges of the veneer as soon as it gels. Scrape the glue off the

Hammer veneering with hot hide glue is fast, cheap, and yields great results with a little practice. This mahogany veneer was glued down in less than five minutes with a veneer hammer, an iron, and a bucket of hot water.

Hammer veneering can be used for precise, delicate work as well as big veneer layups. For small gluing jobs, use a small hammer, like the 6-oz. Warrington-style hammer shown here.

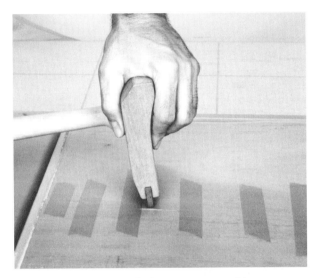

Flat, stable veneers can be hammer veneered in a vacuum press, using the hammer to squeeze out excess glue in the usual manner. This technique allows you to deliver ample pressure to the veneer without using a top caul sheet.

surface once it hardens a bit instead of scrubbing it off with water as it's gelling. As you perfect your personal approach to hammer veneering, remember that it's supposed to be fast and efficient, and it doesn't have to be messy.

Vacuum hammer veneering

Hammering veneered work while it's in a vacuum press is a handy combination of traditional and contemporary veneering techniques. Vacuum hammer veneering allows you to press veneer onto the top face of a substrate without a top caul sheet, which is usually required. The vacuum bag provides enough pressure to position and hold the veneer on the face of the substrate, but not enough pressure to clamp the veneer perfectly flat. You supply the extra pressure that's needed by pushing the veneer down with the veneer hammer, while forcing excess glue from the middle of the panel out to the ends and edges, just as in traditional hammer veneering.

This technique works best with glues that tack quickly, such as PVA. It is most effective on small or narrow panels, because the vacuum bag is only able to hold small veneered surfaces flat once the veneer has been forced down. It also works better with plain-grained veneers and stable species, like mahogany. Crotches and burls can be tough to control, and some veneer species (e.g., oak) swell so much when wet with water-based glue that the bag can't hold them flat, and they wrinkle.

Ironing on veneer with PVA

Like vacuum hammer veneering, ironing on veneer with PVA glue is a handy technique that has limitations. To use this method, coat both the veneer and the substrate with PVA, and let the glue dry (the glue side of the veneer can be sized with shellac first to reduce glue bleed-through if needed). Once the glue has set, the veneer can be ironed onto the substrate. Even though the glue has released most of its moisture, the thermoplastic PVA resins soften under the heat and the glue regains its tack and adhesive properties, much as hot melt does. In fact, if you substitute EVA glue for the PVA with this technique, you're basically veneering with hot melt.

The higher the grade and strength of the glue you use, the sooner you'll need to iron the veneer down and the more heat you'll need to use as you work. With a cross-linking or high-solids industrial PVA, you should iron on the veneer within a day or two. With consumer-grade white glue, you can wait for several weeks if you want to.

When you use this method, you end up with a much thicker and less consistent glue line than usual, which will be less durable than a typical PVA glue line. You can overheat or overwork the PVA resins during the veneering and weaken them if you're not careful. This technique is fine for veneering small or narrow areas, such as an inlaid escutcheon or the edge of a panel, but I wouldn't use it to veneer a large surface such as a tabletop.

Ironing on veneer with aerosol contact cement

Even though contact cement isn't a permanent adhesive, you can use it to bond veneer to a substrate with adequate strength and durability, by using solvent-based aerosol contact cement and an iron. It's a fast, easy technique that I learned from *Fine Woodworking* contributing editor Chris Minick.

Spray the cement on the substrate and veneer (without getting any on the show side of the veneer). Assemble the two after waiting a minute or so, but while the glue is still wet, according to the directions on the can. Then iron down the veneer, applying pressure at the same

Though contact cement is not an ideal veneer glue, you can bond veneer with acceptable strength by using aerosol cement and an iron. Applying heat allows you to press the cement into a thinner layer and improves its penetration.

time. The heat liquefies the cement, increasing its ability to penetrate the gluing surfaces and allowing you to press it into a thinner layer, which increases its strength. Aerosol cement works well for this method because it goes on in thin layers to begin with. Like ironing on veneer with PVA, this technique isn't ideal for large-scale heirloom work, but it's handy for small, less critical applications.

Conditioning

Shop-made substrates and veneered workpieces should be conditioned to allow the large volume of glue they contain to cure. Panels can be stacked, stickered, and top-weighted on a flat surface, which will allow added moisture, heat, and toxic glue vapors to dissipate. Give glued work more time to condition than you would give an edge-glued panel or an assembled framework—several days, at the least. If you stress or machine veneered work before it is properly conditioned, it may bow or twist.

13

Assembly Gluing

Most woodworking projects are made up of dimensioned, shaped, and fitted parts that are assembled into structures, such as cabinets or chairs. These structures are usually glued together, using various joints such as dadoes, dovetails, and mortise-and-tenon joints.

It's important to glue structural assemblies together as well as you can for a couple of reasons. One is that lots of time and effort go into crafting the parts that are being glued together. A poor glue-up can jeopardize or waste all that effort. The second reason is that glued structures encounter all sorts of forces that they have to be able to withstand in order to endure for any length of time.

If you plan and prepare assembly joints carefully and use the right glues and techniques, you'll be able to glue them up properly, and they'll have the best possible chance of enduring once they're glued up.

Stresses That Affect Assemblies

Once a structural assembly is glued up, it can be subject to both internal and external stresses. The internal stresses come from all the mechanical forces that are at work within the glued structure. These forces tend to collect at the joints. For instance, a solid door panel that expands, cups, or twists will stress the mortise-and-tenon joints in the door frame that contains it. When internal forces come from several different directions at once, the strain on a joined structure can be tremendous.

Many structural solid-wood glue joints endure internal stress because of their grain orientation. In edge joints and face laminations, the grain of the glued parts is oriented in the same direction. In structural joints such as bridle joints, though, the joint members are oriented at 90° to each other. As they shrink and

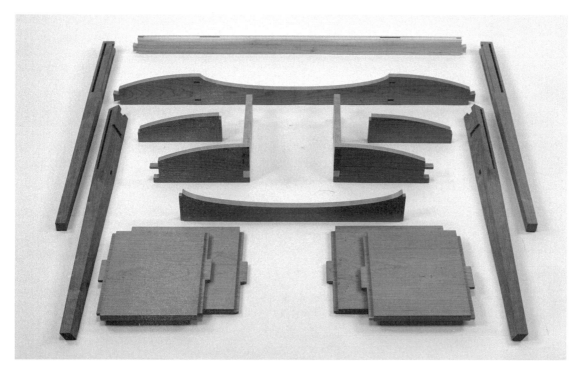

Considering how much time and effort can be spent making the parts of a project (these are for a sideboard by Joe Twichell), it's easy to see why assembly gluing is so critical.

swell in width due to changes in relative humidity, the joint members exert lateral shear forces on the bond layer from perpendicular directions. The useful life of the bond layer depends on how many seasonal cycles of such stress it can withstand before it breaks down.

Bond layers in assembly joints are also subject to opposing lateral shear stresses that are generated by seasonal changes in the thickness of parts. However, in joints like the carcase dovetail joint shown in the photo on p. 144, seasonal changes in the width of the parts won't stress the bond layer because both parts have the same grain orientation and shrink and swell across their width in unison.

Gravity is the chief constant external force that stresses assemblies. The joints in big doors, marble-top tables, and mirror frames all have to carry lots of

Assembly joints have to endure many forms of degrade and abuse. The glue joints in this chair back have been subjected to insect damage, extensive upholstery nailing, hygroscopic wood movement, and seating stresses.

These sunken dovetails were glued with low-grade PVA in a dry shop. The wood gained thickness in normal humidity, and the soft glue couldn't hold the joint flush.

weight relative to their structure and glue surface area. The external forces that are the hardest for assembly joints to handle, though, are produced by people—the ones who slam doors, overload drawers, and tilt backward in chairs.

Joint Design

A good way to help the glue you use keep your work together is to design effective joints. Heft doesn't make good joints—sensible placement and proportioning do. Some of the most stressed joints in furniture history, such as the $5/16$-in.-thick mortise-and-tenon joints found on Colonial Philadelphia chairs, look skimpy but have survived for two-and-a-half centuries because they are sufficiently sized for the job and strategically located within the work.

Detailing is also essential to good joint design. The tenons joining the side rails to the back posts of those Philadelphia chairs run all the way through the back posts and are secured with wedges. Adding details such as wedges, housings, and shoulders to assembly joints will help them endure mechanical stresses without having to depend entirely on glue for all their strength.

Gluing Surfaces

Well-designed assembly joints make good use of effective gluing surfaces. With plywood and other sheet goods, all ends, edges, and faces are acceptable gluing surfaces. In solid wood, however, end grain is not a very effective gluing surface. The best bonds in solid wood come from gluing the long-grain surfaces found on the faces and edges of boards.

Though good joints are based on good gluing surfaces, it's not always wise to use all the gluing surface area available. For example, consider a 12-in.-wide bottom rail on a frame-and-panel door. If you tenoned it across its full width, you'd have a lot of good gluing surface. But the rail's seasonal changes in width would generate lots of lateral tenon movement within the door-stile mortises, which could cause the rail to crack and also cause the bond layer to fail. To help prevent such problems, your best bet would be to divide the full-width rail tenons into two or three smaller tenons and lengthen the mortises slightly to allow room for lateral tenon movement.

Making Joints

There are two goals in assembly joinery: Each joint has to fit properly, and the overall structure has to go together with relative ease. If you have to force a structure together with extra clamping pressure to get all the joints to look tight, you'll add unwanted stress to the assembly, and you'll end up distorting it out of proper alignment as well. Take the time to get things right as you test-fit and dry-clamp your work.

As far as the individual joints themselves go, it's important to think of them as *glue* joints. There has to be enough room in the joints to allow glue to coat the surfaces and bond them properly. Allowing room for glue doesn't have to be complicated. With a mortise and tenon, for example, if you have to wiggle or hammer the joint together when it's dry, it's too tight, and the glue will be pushed off the gluing surfaces when the joint is assembled. If the tenon drops into the mortise with a clunk, it's too loose, and the glue will dry in a thick, fragile layer after assembly. In a well-made joint, the gluing surfaces slide against each other easily, with the wood fibers making consistent light contact as you test-fit the joint.

You don't need to use calipers or feeler gauges to get this kind of fit, but hand tools sure help. I used to spend hours anxiously shimming jigs and tapping machine fences over a whisker as I made joints. Then I learned that sharp chisels and well-tuned shoulder planes do a great job of putting the final touches on joint faces after they have been machined.

If your joints fit well right off the machine, make sure that they weren't burned or glazed during machining, and make sure that they haven't been fouled by grime or lubricants. Don't worry if the joint surfaces aren't dead smooth, though. Carefully handsawn or bandsawn tenon cheeks and mortise walls that have been neatly chopped with a hollow-chisel mortiser make very effective gluing surfaces.

Machine-cut joints don't always fit well or allow the right amount of room for glue. Fine-tuning machined joint surfaces with hand tools greatly improves their fit and their durability once they're glued up.

APPLYING GLUE

When you're gluing assembly joints, you often have to spread small amounts of glue in many different spots, such as the pins and tails of a dovetail joint. Apply glue precisely and quickly so you can have plenty of time to close the assembly and clamp it properly. Specialty applicators will help you get the job done more quickly, and they don't have to be fancy. If you're gluing biscuit joints, for example, just putting a cheap, basic biscuit slot-applicator tip on your glue bottle is a big help. You don't need fancy glue brushes, either. Just make sure they're the right size for the job so you can control the amount of glue you apply.

Apply glue to all surfaces on both halves of a joint, even if they're secondary surfaces such as end grain. Don't be tentative about this; your goal is to do whatever you can to bond that joint indefinitely, and having glue on every surface helps. Be thorough but moderate with the glue, because

Gluing Up on a Reference Board

Gluing up a joined framework is not always as easy as it looks. You can prepare the parts perfectly, cut neat and accurate joints, position your clamps properly, and still end up with a glued frame that's out of square or twisted. To prevent this, you can check the frame for flatness and squareness with winding sticks and a bar gauge as you clamp it up, and reposition the clamps until the frame is properly aligned. Or you can impose flatness and squareness on the frame by gluing it up on a reference board like the one shown in the photos below, which is far less tedious.

To use this method, dry-clamp the frame together, square it up, then clamp the frame rails to the reference board as has been done in the photo below left. Note that the joints of the frame are clear of the reference board, so the frame won't become glued to the board during assembly. Once the rails are clamped to the board, remove the stiles (see the photo below right), apply glue, and reclamp the frame. Keep the frame clamped to the reference board until the assembly clamps are taken off of the frame itself. If you have lots of frames to glue up, use several reference boards (3/4-in. flakeboard is cheap and works fine), so you can glue the frames up and unclamp them in a steady, efficient rotation.

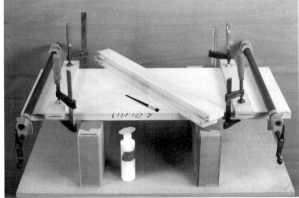

A reference board ensures that a frame assembly will remain square and untwisted. First, the frame is dry-clamped to the board (left), then the stiles are removed (right), so glue can be applied.

excessive squeeze-out can ooze into areas where glue doesn't belong, such as between a solid door panel and its surrounding frame.

If you're gluing miters, or any other joints that depend on end grain as a gluing surface, size or prewet the surfaces with glue just before applying a full glue spread to make sure that the glue will not be soaked up by the porous end grain, which would starve the joint.

CLAMPING

You should be able to close and clamp well-fit assembly joints without heavy clamping pressure once the glue is

Carcases can be glued up simply and accurately on their sides. This cabinet is being supported by shop-made hollow blocks placed on a flat surface. The height of the blocks keeps clamp handles clear of the work table.

Some glue-ups require clever solutions. This custom caul allows the side of the case to be clamped without damaging the beveled leg face. The tape is used as padding.

spread. Excessive clamping pressure can distort an assembly and cause it to dry out of square, or can torque it into a twisted plane. However, some glues tack quickly and swell joint surfaces, which can make it hard to assemble even a well-fit joint. The best strategy is to spread glue and close assemblies as quickly as possible, no matter what kind of glue you're using. If you have to use a lot of clamp pressure to bring a joint home tightly, do so, then loosen the clamp until it's supplying only enough pressure to keep the joint snug while the glue sets. When you clamp assemblies quickly, you allow yourself more time to check them to make sure they're flat and square, and to reposition clamps and cauls as needed to correct the assembly before the glue sets.

Assembly joints can be tricky and time-consuming to clamp up when lots of parts are involved and you have to position special cauls so that they conform to the work and distribute

pressure properly. Keeping things as simple as possible will help glue-ups proceed more swiftly. For instance, you can make dovetails so that the pins will not protrude when the joint is closed. Then you can use a plain block of soft wood as a caul instead of a block that has to be notched to fit around the ends of the pins.

Clean-up

Cleaning excess glue from assembly joints, such as carcase and frame joints, is harder than cleaning up edge or face joints. Most assembly joints include prepared show surfaces that meet at angles and contain corners and other hard-to-clean areas. I prefer to clean up water-based glue from these confined show surfaces while the glue is either still wet or slightly set. It is much harder to remove neatly after it has hardened. Glues such as epoxy and polyurethane glue, which can't be cleaned up with water, are just the opposite. They are difficult to remove completely while they're still wet, but surprisingly easy to clean up once they harden.

If you clean excess water-based glue off an assembly joint while it's still wet, do so thoroughly because traces of glue left on the wood may resist stains and finishes, creating unsightly blotches. In many cases, I don't mind wetting down a glued-up assembly joint to remove excess glue because the water raises the grain of the wood, setting it up nicely so I can sand it smooth with fine sandpaper. I usually incorporate wet glue clean-up into my overall assembly plan by presanding the parts with medium-grit paper (150 grit or 180 grit) before gluing up, then final sanding the glued-up assembly with 220 grit (or finer) paper after I've cleaned off the glue and let the assembly dry.

Besides thoroughness, the key to successful wet clean-up of water-based glue is using the right tools. My clean-up tools aren't expensive, but they are well chosen. I use palette knives, plastic scrapers, inexpensive brushes such as acid swabs and "chip" paint brushes, and thin, durable cotton rags, like those furnished by restaurant-linen supply houses. On clean-up brushes I trim the bristles short and at an angle so I can reach into corners. I also wash out glue rags after use, and they last for a long time.

After gluing up an assembly with a water-based glue, I remove most of the squeeze-out with a plastic scraper or palette knife. Then, using clean hot water (see pp. 103-104), I wet the squeeze-out area with a 1-in. chip brush, wipe the water off with a damp rag, and scrub the remaining glue out of the corners with water and a small acid swab brush. Then I wipe the area with another well-moistened rag, feathering the moisture away from the clean-up area to other areas of the assembly so that the work will have an even appearance after it dries. By using different rags each time, I ensure that no traces of previously removed glue get reapplied to the work. If I want to get the moisture off the work quickly, I blast it with a hair dryer.

Because wet clean-up can be time-consuming, I generally do it only in confined areas of assemblies. To clean up areas that are easily accessible, such as the front and back faces of door frames, I let the squeeze-out partially or completely harden, then scrape it away or pare it off with a chisel before planing the faces flush with a hand plane.

Strengthening Assembly Joints

You can increase the strength and durability of assembly joints by reinforcing them with secondary elements, such as pins and glue blocks. Pins are used mainly with mortise-and-tenon joints. They can be added during or after assembly. They are generally set away from joint lines by about the equivalent of their diameter.

To add pins during assembly, dry-clamp your work, bore holes for the pins through the assembled joints, then dismantle the work. After spreading glue on the joint parts, reassemble and reclamp the work, and drive the pins into the holes before the glue gels.

If you prefer to pin joints in this way, you may want to drawbore your joints. Drawboring involves offsetting the pin holes in the mortise and the tenon slightly (see the top photo at right), so that driving in the pins (which have tapered ends) draws the joints tightly together. If done properly, drawboring eliminates the need for clamps and cauls, which makes gluing up much simpler. It should be practiced on test joints before being used on actual work to determine the right hole size, location, and amount of offset.

If you prefer to pin joints after gluing them up, first decide whether you want to use round or square pins. Round pins are easier to put in; square pins look more traditional and fit tighter because they're pounded into round holes. To install round pins, drill the pin holes anytime after unclamping the assembly. Glue the pins into the holes as soon as you drill them. If you want to use square pins, wait overnight until the glue in the joints has hardened before drilling the holes. Then drive the pins in dry—if you use glue, the pins will be much harder to drive to full depth.

Glue blocks are a quick and easy way to add strength to leg-and-rail joints, carcase joints, and lots of other assemblies. Period cabinetmakers used them wherever they could, including some places where they didn't belong, such as across the grain of certain assemblies. Long, continuous crossgrain glue blocks either fall off or cause the work to split because they restrict seasonal movement. If you add glue blocks crossgrain, cut them into short sections and space them apart so this won't happen.

Though glue blocks should be fit precisely, they don't have to be fancy looking. I make glue blocks out of scrap pine rippings that range in size from ¼ in. to about 1¼ in. square. After cutting blocks to length as needed, I fit

Pinning joints is a great way to strengthen them or to pull them together. On this drawbored joint, note the slight offset of the holes inside the joint. When pins are driven through the holes, the joint will be drawn tight. (Photo by Allan Breed.)

Glue blocks are easy to fit and glue down with rub joints, which don't need to be clamped. The blocks can be trimmed before or after gluing. (Photo by Allan Breed.)

them by truing two adjoining faces with a belt sander and/or block plane until they meet at an angle that matches the angle formed by the gluing surfaces on the work. Then I apply glue, rub them in place to squeeze out the excess glue, and hold them there until the glue hardens enough to hold them fast. I avoid using clamps whenever possible.

Adhesives to Use

No single adhesive is ideal for all assemblies, and there is no specific combination of properties that an assembly glue should have because there are so many different types of assemblies. But there are some general points that are worth considering when choosing an assembly adhesive.

First, the glue should be fresh so it's easy to apply and allows the maximum assembly times it is supposed to allow. Second, the glue should either cling well to gluing surfaces or flow well inside a joint so that all gluing surfaces will be properly coated with glue during assembly and clamping. Third, the glue's degree of initial tack should be well suited to the job at hand. For example, high-tack glue is useful for adding glue blocks, and low-tack glue is desirable for assembling sliding dovetails. Fourth, the glue should allow ample closed assembly

time as well as ample open assembly time because positioning and tightening clamps once the work is closed can take as much time as spreading the glue itself. Finally, the glue's solids content should be composed mainly of adhesive solids such as resins instead of other solids such as fillers and extenders, so that the glue will develop high strength and endurance as it cures.

When you're choosing an assembly glue, consider whether you want it to be easy to repair in the future or not. Assembly joints are often the first glue joints to fail in a woodworking project, and are the joints most likely to need repair. If you assemble joints with an adhesive like urea resin glue, they will be much harder to repair should failure occur. Animal glues are the easiest assembly glues to repair because they are reversible.

Although many adhesives are used for assembly gluing, it's worth looking at which ones are the most useful.

Hide glue is a superb assembly glue, provided that it's used correctly. Its quick, aggressive tack makes it an ideal choice for rubbing on glue blocks and other rapid assembly tasks. Hide glue can also be used for more intricate assemblies if you use a lower-gram-strength glue (see p. 27) or retard the gel rate to allow more assembly time (see p. 29). Hide glue has superior strength and good durability once it's cured.

PVA glue is widely used for assemblies, but few woodworkers employ a variety of PVAs for assembly gluing, as they should. Slow-speed PVAs allow extended open and closed assembly times, which make complex glue-ups much more manageable. At the other end of the spectrum, high-speed

Polyurethane glue is not suitable for biscuit assemblies, even if they are premoistened with water. This joint was broken apart by hand with ease.

assembly PVAs tack and set so quickly that they can be used almost like hide glue or hot melt for rapid assembly work. PVAs are also available in different viscosities. High-viscosity formulas cling well to large joint surfaces; low-viscosity glues flow readily and coat joint surfaces well when joints are assembled. EVA glues cure to a soft, flexible film, which makes them useful for assembly joints that move seasonally, such as breadboard ends on tabletops.

By itself, urea resin glue isn't a great choice for assembly gluing because it requires mixing, it can be brittle, and it's fairly bothersome to clean up, wet or dry. But it can be very effective when added to PVA glue (see the sidebar on p. 73). Urea-fortified PVA is stronger and more rigid than unaltered PVA and has better durability and creep resistance.

Epoxy makes a great assembly adhesive because it's strong and durable, and because it allows long assembly times and fills gaps with true structural strength. However, it's not ideal for all assembly work because it can be hard to work with and clean up. Polyurethane glue is more convenient to use and very easy to clean up once it cures. It's a superb assembly glue but works well only with joints that are well made and closely fit. It is not a good choice for biscuit joints (see the photo on the facing page).

Finally, hot melt is very useful for assembly work, but only as a secondary glue. It isn't strong enough for permanent bonding, but it is the glue of choice for the rapid assembly of temporary fixtures such as jigs and clamping blocks (see the photo above right). On the other hand, the new

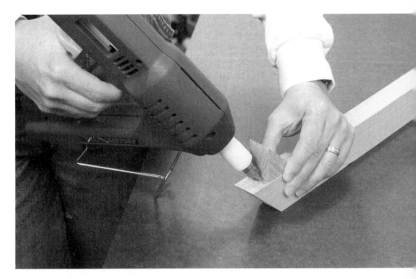

Gluing a temporary clamping block to a mitered frame member with hot melt. A bead of glue in front of and behind the block is all that's needed to hold it in place, and the block is easy to remove after the frame is assembled.

reactive hot melts are very strong permanent adhesives that harden as quickly as regular hot melts. They may well become the premier rapid assembly glues of the future.

Conditioning

A well-fit assembly joint that has been put together with water-based glue will hold moisture for days after the glue-up. If you hand-work or machine an assembly soon after glue-up, there will be problems when the work later releases the added moisture. These problems include puckered biscuit and mortise-and-tenon joints and sunken edge and face joints (see the drawing on p. 115). To avoid such problems, wait at least a day or two before leveling or smoothing the surfaces of glued assemblies to allow the moisture to dissipate.

14

Decorative and Specialty Gluing

Glue does a lot more than hold edge-glued boards and chair frames together. It is also an essential part of many decorative and specialty woodworking techniques and is widely used for repairs. As with many other aspects of woodworking, gluing down decorative elements is not as complicated or tedious as it may appear to be, providing that the right glues are used in the right way. When bonding specialty non-wood materials, such as metal and leather, preparation often plays as big a role as the adhesives that are used. And, of course, there are always repairs. Everyone has to make a repair or two once in a while, whether it's regluing loose veneer on an old dresser or replacing a chip mistakenly gouged out of a just-completed table leg.

Solid-Wood Decorations

One way to enrich your work is to apply solid-wood details to it, such as moldings, split turnings, cock beads, quarter columns and carved leafage. Many of these applied decorations are contoured and hard to clamp. Often the best way to apply them is to ignore clamps and cauls and rub the parts on by hand using hot hide glue, which tacks and gels quickly and forms a strong, durable bond.

The back side of an applied decoration may benefit from a light scoring with a toothing iron to keep it from sliding around a lot as you glue it down and to ensure that it mates snugly and securely with the surface it's being glued to. If you want to rub applied workpieces down with a synthetic glue, the best choice would be a high-speed-assembly PVA glue, which tacks and sets almost as fast as hide glue.

If you're using clamps to glue down applied decorations, hide glue and PVA are still good choices, but you'll need more assembly time so you can get the clamps positioned and tightened. If you're using hide glue, you can add gel depressant, switch to a lower-gram-strength glue, or warm your work before gluing up, all of which extend hide glue's assembly times. If you're using PVA, switch to a regular-speed glue, which allows longer assembly times. Another good choice is polyurethane glue, which takes longer to cure but is easy to clean up after it hardens.

Masking tape is a good alternative to clamps when gluing small solid-wood elements such as sawn stringing strips onto the rabbeted edges of veneered panels. Stretch the masking tape as you pull it over the work. The stretched tape will retract when you let go; this creates enough pressure to hold the stringing in place tightly while the glue sets.

Veneer Decorations

Veneer is a very versatile decorative material. It can be cut and applied in geometric patterns, used for stringing and bandings, and sawn into floral inlay and pictorial marquetry.

When you apply veneer in geometric patterns, say for a radial-matched tabletop or a chessboard, your approach to the job should be determined by the scale of the work and the type of glue you use. Big patterns can be hammer-veneered piece by piece using hide glue or taped together and pressed as a unit. You can use hide glue to press big patterns if you want, provided you warm the work, slow the glue's gel rate down, and heat the work once it's in the press

The contours of applied solid decorations can be difficult to clamp with regular clamps. The clamps on this shell are shop made from bedsprings. (Photo by Allan Breed.)

Masking tape clamps small elements like this corner stringing nicely because it stretches out at application, then rebounds and pulls the work tight.

to reactivate the glue if needed (see the photo on p. 136). Although that method works, most woodworkers opt for a simpler course, which is to press the pattern with a synthetic adhesive such as urea resin glue or PVA. With small-scale patterns, it's best to tape the pieces together and glue them down as a unit, no matter which glue you use.

STRINGING

For single-line string inlay, such as that shown in the photo on p. 157, choose a glue that will tack quickly so you can press the narrow strips of veneer into scratched or routed grooves without using clamps. You can use a small roller to work the inlay into the groove. A small veneer hammer also works well if the stringing isn't fragile or splintery. The main objective in gluing single-line stringing is getting the right amount of glue into the tiny grooves as neatly and efficiently as possible. Otherwise, the job will get messy and take too long. Fish glue, hot hide glue, and PVA are the best choices.

Fish glue works best when the veneer strips are well behaved, but may not tack with enough initial strength to hold unruly strips that have to be coaxed home. It's very convenient to use and easy to clean up, though. Hot hide glue can be convenient to use as well, and it's a better all-around choice. Put the glue in a handy applicator, such as a syringe or a small plastic squeeze bottle fitted with a fine tip, and keep the applicator immersed in hot water so the glue will remain fluid. You can thin the glue slightly if it's hard to squeeze out or retard it if it gels in the groove too quickly. A regular or fast-acting PVA also works well for single-line stringing. PVAs tack fairly quickly, and the moisture they contain will swell an inlaid line and its groove to a tight fit.

With white inlaid lines, you may want to use the lightest-colored glue possible to avoid tinting the inlay. Some batches of regular hide glue are darker than others, but high-clarity hide glue is consistently pale and clear, and is a good choice. Similarly, white PVA is a better choice than tinted PVA, and untinted high-grade PVAs are available if you don't want to use ordinary white glue. To preserve the paleness of white lines and the darkness of black lines cut from dyed veneer, use as little water as possible during clean-up.

MULTI-LINE STRINGING

Multi-line stringing is a decorative element that gets glued twice. First, you laminate a sandwich of contrasting veneers (black and white are the most common), then you glue thinly sawn strips of the lamination to your work. The glue you use to laminate a stringing package should not have an overly high moisture content, or it might cause dark veneers to bleed color into light veneers. It should also be thermoplastic, so you can heat the stringing and coax it around curves if you need to. Multi-line stringing for tight curves can be glued up as a miniature bent lamination.

You can add multi-line stringing to your work in several different ways, depending on the demands of the work. For example, you can plow inlay grooves

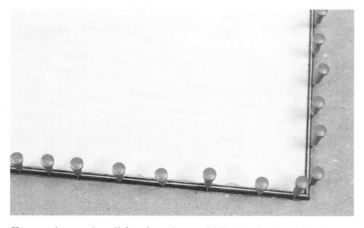

Veneer pins work well for clamping multi-line stringing and other similar decorations against a veneer field or rabbet lip.

for the strips, glue them into rabbets, or glue them against the trimmed edge of a face veneer that has already been glued down. Gluing stringing into a plowed groove proceeds just like single-line inlay. To glue stringing around a shallow lip, you can use pins or the edge of a thin caul that is clamped to the work so that it presses tightly against the stringing. You can also combine both methods, and pin a narrow caul against the stringing. Pinning is a traditional clamping method, and it's very effective. Pins (see the photo on the facing page) are easier to use than cauls when gluing stringing around a curve, but cauls are easier to use along straight sections.

To laminate multi-line stringing, pale or clear hide glue and untinted PVA are the best choices. EVA is a good choice for stringing that will be applied around curves because it can be bent more easily with heat. Even though it's not as strong as PVA, EVA will still hold stringing strips together properly until they can be glued to the work with a stronger adhesive.

To apply strips of multi-line stringing to your work, use the same glues that you would use for single-line stringing. Since gluing multi-line stringing is often more time-consuming than gluing single-line stringing, you may need to retard the glues you're using or switch to slower-speed glues.

ASSEMBLED BANDINGS

Making an intricate banding inlay can take several glue-ups if the core is elaborate and needs to be cut and reassembled repeatedly to create decorative patterns. Whether they're plain or intricate, however, banding cores are composed largely of solid wood, which is arranged so that its end grain is buried in the laminated structure of the banding (see the drawing below). This eliminates visible end grain from the finished banding (which is sawn from the edge of the assembly), but it also packs the banding lamination with porous, inferior gluing surfaces.

The glue you use to laminate a banding log has to be able to bond end grain with some measure of success, so the log will be as sound as possible. Banding strips from a well-glued log should be able to survive being sawn free and glued to your work without coming apart. The glue should also have a natural appearance against wood surfaces, in case it wicks into porous

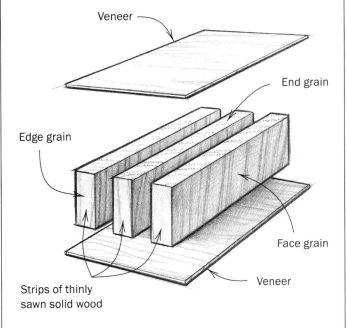

ASSEMBLED BANDING

Veneer

End grain

Edge grain

Face grain

Veneer

Strips of thinly sawn solid wood

Much of the gluing surface inside an assembled banding is composed of end grain, which can be hard to glue.

Vertical-grain crossbanding can be glued in place using spring clamps and Plexiglas cauls. These clamps are cocked back slightly to push the banding in toward the stringing.

core pieces and becomes visible on the faces of banding strips that are sawn from the log. Although fish glue works fine, hide glue is a better choice for banding glue-ups because it's thicker and gels more quickly, allowing you to work at a faster rate. Thicker glue is preferred because it adds less moisture to the banding log (which helps minimize color bleeding between different woods), and won't wick as deeply into the end grain of core parts.

PVA is also a fair choice for banding glue-ups. You can tint PVA so that it will blend in with darker banding woods, and thicken it if needed to control penetration. To glue sawn banding strips to your work, hide glue and PVA are the best choices.

CROSSBANDING AND EDGE BANDING

The borders and edges of panels are often decorated with pieces of vertical-grain veneer, which are known respectively as crossbanding and edge banding. These pieces are short crosscuts taken from wide sheets of veneer. They can be glued down with

clamps and cauls, they can be hammer-veneered onto the work, or they can be ironed down using PVA glue (see pp. 140-141). The ironing method is not ideal for veneering large surfaces, but it works well for smaller areas such as panel borders and edges.

BORDERS

To hammer-veneer crossbanding onto the borders of panels, cut the veneer so that it barely overhangs the edge of the panel, and use a small veneer hammer for more control. Hammer diagonally from the outer edge toward the inner edge of the banding; keep crossgrain hammering to a minimum, or you may split the veneer. If the veneer is fragile, try using a roller instead. When you iron on crossbanding with PVA, don't overheat or overwork the glue, and make sure that the iron doesn't soften any other glued veneers or decorations nearby.

EDGES

Hammer veneering with hide glue and using a hot iron with PVA are also good ways to apply veneer banding to panel edges. Or, using either glue, you can clamp down edge banding with tensioned masking tape as described on p. 153. Before you apply edge banding, size any end-grain gluing surfaces with thinned glue as discussed on p. 130. If you use the masking-tape method, cut the edge banding so that it just slightly exceeds the thickness of the panel edge. Tape won't pull the banding down properly if it projects out from the work too much.

You can use clamps and cauls to glue down crossbanding and edge banding, if you prefer. Crossbanding is a lot easier to clamp than edge banding. I use spring clamps and Plexiglas cauls to glue crossbanding, which allows me to work quickly and see what I'm doing. The easiest way to clamp edge banding is to stand the panel vertically in a bench vise so that the gluing edge is facing up. No matter what technique you use, don't glue down more than 16 to 20 running inches of vertical-grain veneer banding at a time in one area if you are using water-based glue. The banding sections, which are composed of cross-grain material, will shrink in length as they give off added moisture after being glued. If you glue down too long a piece of banding at one time, it will split, crack or separate as it dries out. To make your work efficient, glue short sections of edge banding or crossbanding in several locations around a panel, and rotate succeeding glue-ups between these locations until you've filled in all the spaces between the sections. With a sharp plane and chisel, it's easy to get tight edge joints between the banding sections as you proceed.

Small pieces such as this leaf element of a floral inlay can be either pressed in or clamped, depending on the glue and the quality of the fit. Gluing without clamps makes the work go much faster.

PICTORIAL INLAY

Floral patterns and other veneer decorations are often inlaid in solid wood a piece at a time. The best way to keep this work from becoming tedious and time-consuming is to glue the pieces in without clamps and cauls, as you would with single-line stringing. Once you cut recesses for the elements of the inlay, apply glue to the recess carefully to minimize squeeze-out, and press the element in place. Use a glue that gels quickly and cleans up easily, such as hide glue, or a fast-acting PVA glue. You'll then be able to pare the inlay elements flush soon after gluing them in. Both

hide glue and PVA contain moisture, which will swell the inlay tight in the recesses. If there are any slight gaps in the work, hide glue will dry to a natural appearance as it fills them. If you use PVA, you may want to tint it so that it will blend in with the work if it has to fill gaps.

MARQUETRY

Traditionally, marquetry faces were cut, assembled, and glued down as a unit with hot hide glue, using a weighted and heated caul that would reliquefy the glue so it could bond properly. Hide glue is still a great choice for pressing marquetry faces, if you have a safe, effective way to heat the work. However, most marquetarians today cold-press their work with urea resin glue. Two-part urea resin glue is a better choice than one-part glue. It penetrates better, contains less moisture, and fills small gaps better. It also dries to a natural brown color that can be tinted as needed to blend in with different veneers, but it occasionally stains certain woods like maple and cherry, and it's hard to repair.

To prepare brass strips for inlaying, you can sand the gluing surfaces, as well as treat them with chemical surface preparations and solvents. The phosphoric acid solution shown here and lacquer thinner work well.

Before you press a marquetry face, prepare the glue side by scuffing it lightly with sandpaper. Some of the many different veneers in a marquetry picture may be difficult for glue to penetrate because they contain natural oils, or because their surfaces were compressed during slicing or have oxidized with age. Scuffing freshens and opens up veneer surfaces, which enhances glue penetration. When you press the marquetry onto a substrate, use a rubber sheet between the caul and the veneer. The rubber will equalize the clamping pressure by yielding to variations in the thickness of different veneers and the layer of veneer tape on the show side of the marquetry.

Non-Wood Materials

All woodworkers glue materials other than wood from time to time, usually to bond them to wood. These non-wood materials fall into two categories: rigid materials, such as metal and glass, and pliant materials, such as cloth and paper. Both types of materials can be glued successfully, but they differ in one important respect: Rigid materials generally require some sort of surface preparation before gluing, and pliant materials generally do not.

SURFACE PREPARATION
Metals typically require more gluing surface preparation than other non-wood materials. Each metal has a specific recommended treatment that will produce an optimal bonding surface. These treatments, which are solutions that include ingredients such as mild acids or alkaline salts, prepare bonding surfaces by various means, including etching metal and removing oxide layers. Metal surface preparations are available from auto-body and paint distributors and from suppliers such as Aircraft Spruce & Specialty Company (225 Airport Circle, Corona, CA 91720; 800-824-1930). For detailed information about preparing a specific metal or other non-wood material for bonding, you can contact either the supplier of the material or the supplier of the glue you're going to bond it with.

If you can't diligently prepare non-wood material surfaces for gluing, you should prepare them in some basic fashion so that they can be bonded with a reasonable chance of success. Here are some elementary guidelines to follow:

• **Gluing surfaces should be mechanically clean.** Grease, dirt, scale, and other contaminants should be removed. Rigid or semi-rigid surfaces can be scuffed with sandpaper to key them and expose fresh material.

- **Gluing surfaces should be chemically clean.** Surfaces should be rinsed with distilled water or a solvent such as alcohol as a final step before gluing, then allowed to dry thoroughly.
- **Gluing surfaces should not be touched by bare hands** once they are clean, and should be bonded as soon as possible.

BONDING RIGID NON-WOOD MATERIALS

As a group, rigid non-wood materials have a wide variety of physical characteristics. Bone is porous, has a moisture content, and is acidic. Glass is non-porous and chemically neutral. Iron and steel are chemically affected by contact with wood; plastics are not. Because of these and other differences, you have to treat each material individually when gluing it to wood.

Fish glue and polyurethane glue are the best adhesives to use for bonding metal to wood. Fish glue has been the metal-inlay adhesive of choice for centuries because it wets metal surfaces well and is heat resistant, and it still does the job well. Fish glue is also easy to clean up and repair. Polyurethane glue is a superb glue for metal inlay. It's easy to apply and bonds well, and it is strong, durable, highly resistant to heat, and easy to clean up after it cures, although it's not easy to repair. Heat resistance is essential for metal-inlay glues, because lots of heat can be generated as the inlay is scraped or sanded flush to the surrounding wood surfaces after gluing.

Epoxy is not a good choice for metal inlay because of its low heat resistance, but it bonds metal to wood extremely well where heat isn't a factor. Cyanoacrylate glue is fine for small

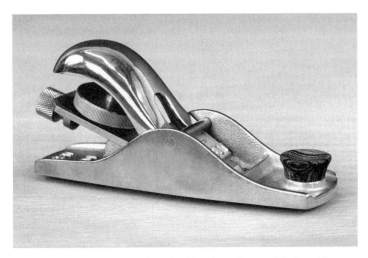

An auxiliary front knob of tropical hardwood was added to this block plane using polyurethane glue, a superb all-around choice for metal-to-wood gluing.

work, but is too fast-acting, expensive, and noxious to use for big metal-to-wood bonds.

Other rigid non-wood materials can be divided into two groups: those that have a workable surface (one that can be easily sawn or sanded) and those that don't. To bond workable materials such as abalone, tagua nut, or plastics to wood, the best glues to use are fish glue, hide glue, EVA, polyurethane glue, and cyanoacrylate glue. To bond non-workable materials such as porcelain and glass, your best choices are epoxy, polyurethane glue, and cyanoacrylate. When working with clear materials, choose an adhesive that has good optical clarity, such as a specialty cyanoacrylate or epoxy. Make sure the adhesive is also non-yellowing, because many clear adhesives yellow with age.

BONDING PLIANT NON-WOOD MATERIALS

Many pliant non-wood materials, such as cloth or leather, are porous and can be bonded with most wood glues. Many of

Repairs lead you down gluing and clamping paths that you'd otherwise never get to travel. This mirror frame is being repaired with rabbit-skin glue.

these materials are also hygroscopic and will shrink and swell along with solid wood in response to seasonal humidity changes. But they can also attract and hold moisture, which may degrade or ruin some glue bonds. Don't glue hygroscopic materials with glues such as simple rice or wheat pastes, fish glue, or clear polyvinyl alcohol craft glue. Instead, use rice or wheat pastes that have antibacterial additives or add some hide glue to fish glue to give it a bit more moisture resistance.

Hot hide glue by itself is usually fine. Both fish glue and hide glue can be plasticized with glycerine if their cured bonds are too rigid for the work. Consumer-grade white and yellow PVAs and EVAs are also suitable for gluing pliant materials.

If you use a water-based glue, make sure it has a low moisture content if possible, so it won't cause porous pliant materials to buckle or stretch. If a glue's viscosity is too low, you can thicken it with an inert additive such as colloidal silica to prevent bleed-through. You can also use heat to speed-cure PVAs and some contact cements once the work is assembled, so the glue will harden before it has a chance to bleed through.

Repairs

Furniture repairs often include tending to glued wood that has failed for one reason or another. Although in most cases, the glue in failed joints is in poor shape, there are some instances where the existing glue either can be repaired or has to be repaired because the joint can't be taken apart. The chart on the facing page gives some basic guidelines

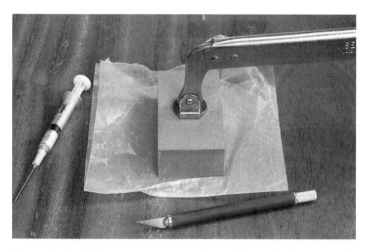

To repair a bubble in a newly veneered panel face, slice it open with the grain, inject fresh glue, and clamp flat.

Repairing Glue Joints

Original Glue Used	Repair Glue(s)	Methods of Repair
Fish glue	Fish glue, hide glue	Reactivate glue with water, add more glue if needed. or Remove old glue, clean joint, reglue.
Hide glue	Fish glue, hide glue	Reactivate glue with water and heat, add more glue if needed. or Remove old glue, clean joint, reglue.
PVA	Freshly glued work: PVA, hide glue, fish glue. Older work: epoxy, hide glue, fish glue	Freshly glued work: Apply heat to soften glue, add more glue if needed, reassemble with heat. Older work: Remove old glue, clean joint, reglue.
Contact cement	Contact cement	Reactivate cement with toluol. or Add new cement, then press or clamp.
Hot melt	Hot melt	Remove old glue and reglue. or Reheat area, add new glue as needed, reassemble.
Urea and resorcinol resin glues	Urea resin glue, resorcinol, epoxy	Remove old glue and clean joint as possible, scuff with sandpaper, reglue.
Epoxy	Epoxy	Remove old epoxy, clean joint as possible, scuff with sandpaper, reglue.
Polyurethane glue	Polyurethane glue, epoxy	Remove old glue and clean joint as possible, scuff with sandpaper, reglue.
Cyanoacrylate	Cyanoacrylate	Remove old glue as possible. Scuff with sandpaper, reglue. Avoid using accelerator.

Resourceful gluing is often a blend of old and new. Here a tabletop is repaired using a go-bar, fish glue, and a Plexiglas caul.

for repairing existing glue layers, but because every repair situation is different, there is no guarantee that these methods will work every time you use them.

CLEANING OFF OLD GLUE

When repairing glue joints, I usually try to remove the old glue as thoroughly as I can. Old glue prevents fresh glue from adhering properly to the joint surfaces. Natural glues are reversible and can be reactivated with moisture or removed with heat and water. Hide glue can also be crystallized with alcohol, which makes the glue brittle and fragile

and allows joints to be shocked apart. This technique doesn't work well unless the glue is pretty well desiccated to begin with. I don't use it much because I'm usually trying to preserve the finish on a piece and can't afford to have alcohol leaking out of joints and streaming onto show surfaces.

The easiest synthetic glue to remove is consumer-grade white PVA glue. It actually remains reversible with water for several months after a glue-up and can be soaked off easily with hot water. Yellow PVA glue becomes insoluble much more quickly and dries to a harder, tougher film as well. Once PVA becomes insoluble, there is no single good way to remove it—you have to use a variety of tactics. I use heat, hot water, acetic acid (or white vinegar), and an assortment of knives, picks, and scrapers. The goal is to soften the PVA film so that it can be mechanically removed without disturbing the joint surface beneath it.

Hot melt can be pared off with a sharp tool, softened with heat and scraped off, or cleaned off with solvents such as acetone, toluene, or naphtha. Some epoxies are easy to remove with heat or acetone, which soften the adhesive film so it can be pared or scraped off. Other synthetic adhesives are not easy to remove. Contact cement is simply a mess, whether you remove it mechanically or with solvents such as toluene or naphtha. Polyurethane glue that has cured to a thin, hard film has to be scraped or sanded off, as do urea and resorcinol resin glues. Cyanoacrylate can sometimes be tapped off with a hammer because it has low shock resistance, or it can be dissolved with debonder.

CHOOSING REPAIR GLUES

Sixty years ago, regluing furniture was simple and straightforward. In most cases, hide glue was used to fix an object that had originally been assembled with hide glue. In the decades since, most of the new adhesives that have been introduced have been used as repair glues.

The good news is that animal glues (hide glue and fish glue) are still the best repair adhesives available. They bond well to older joint surfaces, which often are not pristine. They're compatible with traces of animal glue that may remain inside a repair joint. They blend in visually with older wooden surfaces, and they're very compatible with finishes. Most of all, though, they're reversible (see the sidebar on p. 22), which makes any future repairs easier on both the object and the restorer.

The bad news is that the non-reversible synthetic glues that have been used for repairs over the past six decades have made a mess of many pieces of furniture. Most synthetic glues don't bond very well to old joint surfaces, but can still be difficult to remove from those surfaces. Joints often can't be cleaned properly for regluing without altering the joint surfaces themselves. This ruins the joints' chances of being reglued with strength and endurance. Whenever a non-reversible synthetic glue is used to repair an object, all future repairs become more difficult to do and can be harmful to the integrity of the piece.

Sometimes, though, synthetics have to be used to make repairs. Usually, this is because a previous repair made with a synthetic glue has compromised an object's structure. For example, a joint that was originally glued with hide glue may have been poorly repaired with

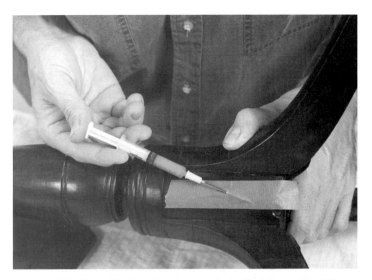

The crack in this table base has been sealed off with tape. As hide glue is injected under the tape, it is forced deep into the crack. When the crack is full of glue, the tape will be removed and the crack clamped shut.

PVA. If it comes apart again (which happens a lot), the joint may be so fragmented or sloppy that the only practical alternative is repair it with a specialty epoxy that will fill gaps and return some structural strength and endurance to the object. Epoxy is not an ideal repair adhesive, but it can often be removed if needed with heat and solvents.

Adhesives are also used to consolidate wood surfaces that have been damaged by insects, rot, upholstery nailing, or repeated breaks. Animal glues can't always be used for these repairs; epoxy may be needed for drastic structural consolidation. For areas riddled with upholstery-nail holes, EVA makes a good consolidant, either by itself or mixed with other glues, because it's reversible and because it's soft when cured and will take further upholstery nailing.

Glossary

Accelerator: a chemical agent that speeds up an adhesive's reaction rate.

Adherend: a material that an adhesive is applied to and will adhere to.

Adhesion: the attachment between an adhesive and an adherend. There are two types: mechanical adhesion (a physical link that develops as glue penetrates into wood pores and hardens, enabling it to grasp the wood) and specific adhesion (a molecular bond that develops between adhesive and adherend).

Adhesive: 1) in general, a substance that forms bonds with an adherend. A widely used synonym for glue. 2) in strict terms, a bonding agent synthesized by chemical means.

Aliphatic resin: a term coined to classify yellow PVA glues and distinguish them from white glues. However, the phrase "aliphatic resin" is a generic chemical description that can be applied to all PVA glues, regardless of color.

Assembly: 1) any two or more workpieces that have been joined together. 2) in industry, gluing crafted workpieces into a structural framework.

Assembly joint: a type of joint used in a structural framework, such as a dovetail.

Assembly time: the period of effective working time that an adhesive allows before it loses its fluidity and impedes assembly or impairs the quality of a cured bond layer. Open assembly time is the period from the beginning of glue application to the moment when an assembly is closed, or fitted together. Closed assembly time is the period from the initial closing of an assembly to the application of full clamping pressure.

Backmelt: the accidental flow of hot melt out of the infeed port of a hot-glue gun.

Banding: 1) a decorative inlay made of laminated solid wood and veneers. 2) veneers that are glued to the borders and edges of panels or frames.

Binder: a component added to a formula to enhance or provide cohesion.

Bleed-through: the seeping of glue through the pores of a veneer, either during glue application or while the veneer is being pressed onto a substrate.

Blotching: a variation in the appearance of finished work caused by traces of excess glue left on the surface that resist stains and top coats.

Bond: noun: the union between an adherend and an applied adhesive that has cured. verb: to join adherends with an applied adhesive layer between them.

Bond line: *See* Glue line.

Bricklaying: a method of constructing curved solid substrates for veneering from short pieces of wood that are glued together in staggered courses.

CA: an abbreviation for cyanoacrylate glue.

Carrier: the liquid medium that an adhesive solid is dissolved or dispersed in. Also known as a vehicle.

Catalyst: an additive that induces or enhances a chemical reaction, but which does not undergo a change itself or take part in the reaction.

Caul: a piece of wood or other material placed between the work and the clamps.

Cement: an adhesive that dries to a film with good adhesion and poor cohesion.

CFC: an abbreviation for chlorofluoro-carbons, a class of chemicals.

Chalk temperature: the temperature below which an adhesive will chalk out.

Chalking (out): a condition that occurs in PVA glue that has cured at too low a temperature. Chalked glue turns whitish and produces poor-quality bonds.

Clamp time: the period of time during which glued work is required to remain under clamping pressure. Also called pressure period or press time.

Closed assembly time: *See* Assembly time.

Cohesion: an adhesive's ability to form internal molecular bonds.

Cold flow: *See* Creep.

Cold pressing: pressing panels or veneers in room-temperature conditions.

Collagen: the protein from which the polymer molecules in animal glues are formed.

Colloid: a solution, such as hide glue, that contains dispersed particles of a solid that are larger than the solution molecules themselves.

Conditioning: 1) putting freshly glued work aside for a period of time to allow the glue to cure and to allow the work to release moisture, heat, and fumes. 2) applying a solution to veneer to make it more workable.

Conditioner: 1) a solution typically made from glycerine, water, and alcohol that is applied to veneer to make it flatter and more flexible. 2) a thinned glue, usually hide glue, that is applied to wood surfaces prior to finishing to make the surfaces accept stains and finishes uniformly, which prevents blotching and other problems.

Copolymer: a polymer that is formed from more than one type of monomer.

Crazing: the formation of cracks and fissures in a cured bond layer.

Creep: noticeable shifting or other movement of wood along a glue joint that the glue is unable to restrict. Also called cold flow.

Crossbanding: 1) vertical-grain veneer glued to the borders of a panel face. 2) intermediate veneer layers in a panel whose grain runs perpendicular to the grain of the core and face plies.

Cross-link: a molecular bond that develops between the polymer strands in an adhesive during the curing process that increases its durability.

Cure: the conversion of an adhesive from a fluid, workable material into a solid. Initial cure is the stage at which an adhesive has hardened, but has not yet developed its ultimate properties. Final cure is the stage at which an adhesive has developed its ultimate properties.

Cure period (cure time): the time it takes an adhesive to cure. *See* Cure.

Cure speed (cure rate): the rate at which an adhesive cures. *See* Cure.

Debonder: a solvent that will soften and dissolve hardened cyanoacrylate glue, allowing bonded objects to be taken apart.

Delamination: the separation of glued elements in a workpiece due to failure of the bond layer between them.

Depressant: an additive that retards the gel rate of hide glue. *See also* Retarder.

Desiccated: dried to a point of degrade.

Diluent: a functional liquid component of a resin formula that is added to reduce its viscosity in lieu of a solvent.

Dimensional stability: the ability of a material or an assembly to maintain its dimensions under variable conditions.

Dispersion: a solution containing solid particles that are permanently suspended.

Dry strength: the strength of a fully cured adhesive before it has been subjected to stress, exposure to variable conditions, aging, and other forces.

Edge banding: a veneer or overlay glued to the edges of panels and cabinet parts.

Edge gluing: gluing the long edge of a solid plank, a panel, or a lamination to another workpiece.

Elasticity: the ability of a material to return to its original shape, dimension, or condition after being stressed or deformed.

Elastomer: an adhesive, sealant, or other material with elastic properties.

Emulsion: a liquid mixture that contains small dispersed droplets of a second, immiscible liquid.

EVA: an abbreviation for ethylene vinyl acetate glue.

Evaporative: having the ability to cure by releasing moisture or solvents.

Exothermic: the ability of a mixture to produce heat during a chemical reaction.

Extender: a semi-adhesive or non-adhesive substance that is added to an adhesive mixture to allow it to coat a larger surface area.

Filler: a non-adhesive substance that is added to an adhesive to enhance its per-formance and/or cured properties.

Flattener: a solution typically made from glycerine, water, alcohol, and glue that conditions veneer and keeps it flat after conditioning.

Flow: the ability of an adhesive to main-tain its fluidity within a closed assembly once pressure has been applied.

Foam-out: the expanding, bubbly excess created as polyurethane glue reacts with moisture while it cures.

Fortifier: an additive that enhances an adhesive's ultimate properties.

Freeze/thaw stability: the ability of a glue to withstand one or more cycles of freezing and thawing without deteriorating.

Gap filling: the ability of an adhesive to fill surface and structural gaps as it cures.

Gel rate: the rate at which a glue converts from a liquid to a semi-solid state.

Glue: 1) a widely used synonym for adhesive. 2) in strict terms, a bonding agent derived from natural substances, such as animal protein or plant starch.

Glue line: the layer of adhesive between two adherend surfaces that has cured to a solid film after application, assembly, and clamping.

Go-bars: lengths of straight-grained wood that exert a clamping force when sprung between a workpiece and an overhead structure or surface.

Gram strength: the measure of a hide glue's molecular weight. Glues with high-er gram strengths are called high-test glues, and glues with lower gram strengths are called low-test glues.

Green strength: *See* Initial tack.

Hammer veneering: a traditional method of gluing veneer using hot hide glue and a squeegee-like veneer "hammer" instead of clamps and cauls or a press.

Hardener: a substance that is combined with a resin to create and control a poly-merization reaction. The hardener takes part in the reaction and becomes part of the resulting cured solid, as with epoxy.

HDT: an abbreviation for heat deflection temperature, which is the temperature at which a thermoplastic material will begin to soften and yield to a load.

Homopolymer: a polymer formed from only one type of monomer.

Honeycomb: an open-grid sheet mate-rial used to make hollow-core panels.

Hot pressing: pressing glued work in a dedicated hot press.

Hygroscopic: having the ability to exchange moisture with the surrounding atmosphere and undergo changes as a result.

Immiscibility: the inability of one liquid to mix with another liquid.

Initial strength: the amount of strength an adhesive develops at initial cure. *See also* Cure.

Initial tack: the adhesive and cohesive strength of wet glue coatings on joint sur-faces at initial assembly. Also called green strength.

Lamina: an individual layer in a lamination (plural: laminae).

Laminate: noun: a layered material that is glue bonded or glue impregnated. verb: to build up a workpiece from glued layers.

Lay-up: glue coating and assembling panel materials for pressing.

Load: a force that is applied to a material or a structure.

Mastic: a highly viscous adhesive material that is used in thick layers and has some gap-filling ability.

Matrix: the body of an adhesive mix that contains suspended additives, such as silica and glass fibers.

MC: an abbreviation for moisture content. *See* Moisture content.

MDF: an abbreviation for medium-density fiberboard.

Mechanical removal: removal of adhesive from a surface with tools or abrasives.

Melt point: the temperature at which a hot-melt glue becomes fluid.

Miscibility: the ability of one liquid to mix with another liquid.

Mix ratio: the recommended ratio of parts in an adhesive batch mix.

Modifier: an additive that alters one or more of the properties of an adhesive.

Moisture content: 1) the amount of moisture contained in a porous workpiece. 2) the amount of water that an adhesive formula contains.

Moisture resistance: the ability of a cured adhesive film to withstand exposure to various specified levels of moisture.

Monomer: a molecule that bonds with other monomers to create polymers.

MSDS: an abbreviation for Material Safety Data Sheet. MSDS sheets list the hazards involved in handling, storing, and using a product.

MSGL: an abbreviation for thousand feet of single glue line, a measure used in specifying adhesive spread thicknesses. *See also* Spread thickness.

Neoprene: a synthetic rubber that is the basis for many contact-cement formulas.

Off-gassing: the release of vapors from glued work after the glue has hardened. Work glued with urea resin glue off-gasses formaldehyde fumes, for example.

Off-ratio: a mix ratio that varies from the specified mix ratio.

One-part: an adhesive whose main ingredients are blended together in one component. One-part glues can be ready to use, such as PVA, or require mixing with water, such as powdered urea resin ("plastic resin") glue or casein glue.

Open assembly time: *See* Assembly time.

Overlay: a decorative and/or protective panel face made from metal foil, resin-impregnated paper, or other materials.

Panel: any workpiece with a relatively large surface area, such as a tabletop or a drawer bottom, and made from either solid wood or sheet materials.

Penetration: the ability of an adhesive to permeate an adherend surface.

Permanent adhesive: an adhesive with a cured strength that exceeds the shear strength of wood, parallel to the grain.

Phenolic: a type of resin used in the manufacture of adhesives such as resorcinol.

Plastic resin glue: a generic name for one-part powdered urea-formaldehyde glue.

Plasticity: the degree of flexibility in a cured adhesive film that will allow it to yield to a force while maintaining its integrity.

Plasticizer: an additive that lowers a glue's rigidity and increases its plasticity.

Platen: a large, flat surface that glued work is pressed against in a veneer press.

Polymer: a large molecule made up of smaller monomer molecules.

Polymerization: a chemical reaction in which polymer molecules are formed from monomer molecules.

Porosity: the degree to which an adherend surface is porous.

Post cure: the heating of glued work after an adhesive has reached initial cure to speed its rate of final cure and enhance its ultimate properties.

Pot life: the period of time during which a batch mix will remain usable if it is kept in the mix pot.

Precatalyzed: describes a one-part adhesive whose catalyst has already been added.

Prewetting: coating a porous gluing surface with adhesive before applying a full-spread thickness of the same adhesive to prevent starvation of the bond layer.

Precure: the premature hardening of an adhesive, either in storage or during use.

Press time: *See* Clamp time.

Pressure period: *See* Clamp time.

PSA: an abbreviation for pressure-sensitive adhesive.

Pucker: a surface depression that occurs in mortise-and-tenon and in biscuit joints. *See also* Sunken joint.

PVA: an abbreviation for polyvinyl acetate glue.

Reactive: having the ability to interact chemically.

Resin: 1) a chemically derived solid that is the foundation of synthetic adhesives. 2) a natural substance found in certain species of wood.

Retarder: a substance added to an adhesive to slow down the rate at which it sets and cures

Reversibility: the ability of an adhesive to be returned to a fluid state after it has hardened, either by heat, a solvent, or both.

Room-temperature adhesive: an adhesive that hardens at temperatures ranging from 65°F to 85°F.

Rub joint: a glued assembly that is closed and clamped entirely with hand pressure, which is applied while rubbing the glued workpieces together.

Sandability: the ease with which a hardened adhesive can be sanded.

Scuff: to abrade a surface lightly and evenly with sandpaper.

Set: the initial hardening of an adhesive once it has been applied. *See* Cure.

Setting speed: the rate at which an adhesive sets.

Shear strength: the measure of a glue's adhesive strength in comparison to the shear strength of wood, parallel to the grain.

Shelf life: the expected useful life of an adhesive product, from the time of manufacture to the point of spoilage.

Shrinkage: the loss of volume that an adhesive layer undergoes as it cures.

Silica: a powdery, inert additive that is used as a thickener.

Size: a preliminary glue coating that is applied to a gluing surface before the full spread thickness of glue to reduce absorbency and improve bonding. Unlike a prewet coat, a size is usually left to dry.

Skin: a thin face ply that is bonded to the core of a panel.

Solids content: the percentage by weight of the adhesive and non-adhesive solids in an adhesive formula.

Solubility: the ability of a solid to be dissolved in a liquid.

Spread: the applied wet adhesive coating on an adherend surface.

Spread thickness: the specified thickness of an adhesive spread. Thicknesses are given in terms of weight of glue per surface area, or lb./MSGL.

Spring joint: an edge joint whose halves are hand-planed to a slightly concave shape along their lengths so that they touch only at each end when pressure is applied but mate snugly once they are clamped together.

Squeeze-out: excess glue that is forced out of joints during closing and clamping.

Starved joint: a joint that does not contain enough cured adhesive in the glue line, due to insufficient spread thickness, overabsorption, or overclamping.

Strength: the ability of a cured adhesive to resist various types of applied stress, such as shear, tension, compression, torsion, and flexion.

Stringing: decorative inlaid veneer strips.

Substrate: an underlying material that a veneer or overlay is bonded to.

Sunken joint: a joint that is leveled flush while it retains moisture added during gluing and that recedes below the flush plane when the moisture dissipates.

Surface sensitivity: an adhesive's sensitivity to the conditions found on a gluing surface, such as moisture, contaminants, and pH.

Suspension: a heterogeneous mixture containing solid particles that settle out when the mixture is at rest.

Synthetic: a substance synthesized by means of a chemical reaction.

Tack: the ability of a wet adhesive to adhere on contact to a gluing surface with a measurable degree of strength. *See also* Initial tack.

Tackifier: a substance added to an adhesive formula to increase its tack.

Telegraphing: the transmitting of defects or other aspects from within a panel core to its surface, resulting in visible variations. Also called showthrough.

Tensile strength: the cohesive strength of an adhesive, as measured by its ability to withstand applied tension. *See* Strength.

Thermoplastic: having the ability to soften and become pliant if heated.

Thermosetting: referring to adhesives whose curing either requires heat or is enhanced by heat.

Thickener: a substance added to an adhesive batch mix to increase its viscosity.

Thixotropic: the ability of a substance to thicken when at rest and flow when manipulated. Certain thickeners can be added to an adhesive to give it this property.

Toothing: scoring a substrate surface with a grooved toothing iron (either hand held or set in a plane) in preparation for veneering.

Torsion box: a hollow-core panel comprised of a wooden grid faced with thin sheet materials, such as plywood. Also called a stressed-skin panel.

UF: an abbreviation for urea-formaldehyde resin glue.

Ultimate properties: the level of physical properties that an adhesive possesses at final cure.

Up-stabilized: refers to cyanoacrylate glue that has been modified with a large amount of stabilizer.

Urea: an organic compound that is used as a gel depressant in hide glue and as a reactive component, along with formaldehyde, in urea resin glue.

Vehicle: *See* Carrier.

Viscosity: the resistance to flow of a fluid substance; in basic terms, its thickness.

Walnut-shell flour: an additive that is used as an extender or to add color and desired consistency to a glue mixture.

Waterproof: refers to the ability of a cured adhesive to withstand continual immersion in water at extreme temperatures without deterioration.

Water resistant: refers to the ability of a cured adhesive to endure intermittent exposure to various levels of moisture at moderate temperatures without degrade.

Wet out: *See* Prewetting.

Wet strength: the shear strength of a cured glue line that is subjected to moisture.

Wetability: the ability of a gluing surface to absorb a liquid adhesive readily.

Wetting: the thorough dispersion of a wet adhesive film on an adherend surface.

Wood flour: an additive that is used as a thickener or an extender.

Workability: the ability of a wet adhesive to be applied with ease and speed.

Working life: an adhesive's period of usability once it has been mixed and dispensed from a mix pot. Working life is not the same as pot life, as some glues remain fluid longer outside the mix pot. *See also* Pot life.

Working temperature: the temperature that a glue requires to cure properly.

Working time: *See* Assembly time.

Index

A

Abalone, glues for, 159
Adhesion:
  by chemical conversion, 9
  by evaporation, 9
  mechanical vs. specific, 8
  by thermal conversion, 9-10
Aliphatic resin, defined, 34
Applicators:
  cleaning, 109
  types of, 109

B

Bandings, gluing advice for,
  155-157
Bent laminations, 121-23
Bone glue:
  conversion properties of, 10
  discussed, 31

C

Casein glue:
  conversion properties of, 10
  discussed, 31-32
  formulation of, 32
  handling and storage of,
    32-33
  performance of, 33
Cauls, sprung, 111
Chalking. See PVA glue.
Clamp time, as factor in gluing
  success, 112
Clamping blocks, 151
Clamping:
  of applied decorations, 153
  with bedsprings, 153
  of crossbanding, 156-57
  of edge banding, 156-57
  as factor in gluing success,
    111-13
  with masking tape, 153
  shop setup for, 104
  of stringing, 153
Clamps, tune-up for, 104
Clean-up:
  of assembly joints, 147-48
  discussed, 13-14, 113
  See also specific glues.
Cloth, gluing advice for,
  159-60
Cold creep. See Creep.
Cold flow. See Creep.
Conditioning:
  of glued work, 114-15
  of joints, 151
  of veneered work, 141
Contact cement,
  aerosol, 48
  application of, 50-52
  clean-up of, 49-50
  conversion properties of, 10
  discussed, 45-46
  drying time of, 51
  endurance qualities of, 52
  formulation of, 46
  grades of, 46

and health/safety issues, 50
for iron-on veneering, 141
for plastic laminate, 137
rollers and spray guns for,
  50, 51
shelf life of, 48
solids percentage of, 48
storage of, 48-49
structural qualities of, 52
thinning, 49
tinted, 46-47
types of, 46, 47
Creep, 106
Crosslinking, 10, 11
Cyanoacrylate glue:
  accelerators for, 93-94
  application of, 99
  applicators for, 97
  assembly times of, 99
  clean-up and disposal of, 98
  color of, 92-93
  containers for, 97
  conversion properties of, 10
  cured working qualities of,
    100
  debonders for, 94, 95
  described, 91-92
  as gap filler, 100
  grades of, 92
  and health/safety issues,
    98-99
  odor of, 93
  removing, 162
  for rigid non-wood materials,
    159
  shelf life of, 96
  stabilizers for, 93
  storage of, 96
  structural qualities of, 100
  surface sensitivity of, 94-95
  thickeners for, 93
  types of, 92
  viscosity of, 93

D

Decorations, applied, gluing
  advice for, 152-53

E

Edges, jointing, 118
Epoxy:
  assembly times of, 83
  clamping pressure with, 83
  clean-up of, 80-81
  color of, 75
  composition of, 74
  conversion properties of, 10
  as crack filler, 77
  cured working qualities of, 84
  cure speed of, 83-84
  for edge and face joints, 121
  endurance qualities of, 84
  "five-minute," 75-76
  as gap filler, 84
  glue-line thickness of, 83
  grades of, 75
  hardeners for, 75

and health/safety issues,
  81-82
for joints, 151
mixing, 77-79
moisture content of, 77
pot life of, 80, 81
pump dispensers for, 78
for rigid non-wood
  materials, 159
shelf life of, 77
for structural consolidation,
  163
structural qualities of, 84
surface sensitivity of, 76-77
thinning, 79
tinting, 79
types of, 76
for veneering, 138
viscosity of, 75
working temperatures for,
  82-83
EVA glue:
  conversion properties of, 10
  discussed, 38
  for rigid non-wood materials,
    159
  for stringing, 155
  for structural consolidation,
    163

F

Fish glue:
  conversion properties of, 10
  discussed, 23-25
  for rigid non-wood materials,
    159
  for stringing, 154
Flakeboard. See Sheet goods.
"Foam-out." See Polyurethane
  glue.
Formulation, 10-14
  See also specific glues.
Freeze/thaw stability, 13
  See also specific glues.

G

Glass, glues for, 159
Glue blocks, 149
Glue pots, 29
Glue:
  adhesive ratings of, 17
  applying, 145-46
  conversion properties of, 10
  fluorescing agent for, 13
  as gap filler, 18, 19
  natural vs. synthetic, 20
  permanence of, 19.
  plant vs. animal, 20-21
  removing, 162
  stirrers for, 40
  storage of, 104
  strength of, 18-19
  See also specific glues.
Glue-ups:
  preparation for, 108
  stresses in, 142-44
  See also Joints.

H

Health, information on, 14
Heat:
  for bent laminations, 123
  as factor in gluing success,
    112, 113
  low-tech setup for, 112-13
Hide glue:
  for applied decorations, 153
  assembly times of, 28-30
  for bent laminations, 122
  for bricklaid curved
    substrates, 132
  clamp time of, 30
  discussed, 25-26
  for edge joints, 120
  endurance qualities of, 31
  formulation of, 26-27
  gel depressant for, 29
  glue pot for, 29
  gram strength of, 26-27
  heating equipment for, 28
  for inlay, 157
  for joints, 150
  for marquetry, 157
  preparation of, 27
  for rigid non-wood
    materials, 159
  for stringing, 154, 155
  structural qualities, of, 31
  for veneering, 136
Hot melt:
  assembly time of, 58
  clean-up of, 57
  conversion properties of, 10
  cured working qualities of, 59
  described, 53-54
  endurance qualities of, 59
  glue guns for, 54, 56
  grades of, 55
  and health/safety issues, 57
  for joints, 151
  melt point of, 56
  for plastic laminate, 137
  removing, 162
  shelf life of, 57
  stick sizes of, 57
  structural qualities of, 58-59
  surface sensitivity of, 56
  types of, 54-55
  and working temperature, 58
Humidity, as factor in gluing
  success, 102-103

I

Inlay, gluing advice for, 157

J

Joint design, as factor in gluing
  success, 144
Joints:
  assembly,
    as factor in gluing success,
      144-45
    cauls for, 147
    clamping, 146-47

cleaning glue from,
147-48
glues for, 150-51
preparing, for glue-up,
106-107
reinforcement for, 148-49
edge,
glues for, 120-21
keyed, 117
sprung, 118
with sheet goods, 124-25
end,
glues for, 121
gluing advice for, 119
face,
glues for, 120-21
gluing advice for, 119
with sheet goods, 124-25
puckered, 115
sunken, 115

L

Leather, gluing advice for,
159-60
Light, as factor in gluing
success, 103

M

Marquetry, gluing advice for,
157-58
MDF. *See* Sheet goods.
Metal, gluing advice for,
158-59
Moisture content, discussed, 12
MSGL, conversion formula for,
110

P

Penetration, as factor in gluing
success, 110
Performance, 14-18
See also specific glues.
Phenol-resorcinol glue. *See*
Resorcinol resin glue.
Pins, as joint reinforcement,
148-49
Planes, toothing, 129-30
Plastic, glues for, 159
Plastic laminate, glues for,
137-38
Plywood. *See* Sheet goods.
Polymers:
adhesive, 7
bonding of, 8-9
crosslinking of, 11
Polyurethane glue:
accelerator for, 89
application of, 88-89
assembly times of, 89
for bent laminations, 123
for biscuit joints, disadvised,
150
clamp time with, 89
clamping pressure with, 89
clean-up of, 87-88
composition of, 85
conversion properties of, 10
curing requirements of, 90
development of, 85
for edge and face joints, 121
for lumber-core plywood
substrates, 132

"foam-out" of, 86, 90
formulation of, 86-87
and health/safety issues, 88
for rigid non-wood materials,
159
storage of, 87
structural qualities of, 90
for torsion-box substrates,
133
for veneering, 138
Porcelain, glues for, 159
Pot life, 13
See also specific glues.
PVA glue:
for applied decorations,
152-53
assembly times for, 42-43
for bandings, 156
for bent laminations, 122-23
for bricklaid curved
substrates, 132
catalyzed, 35
chalking with, 42
clean-up of, 40-41
composition of, 36
consumer- vs. industrial-
grade, 34-35
conversion properties of, 10
cured working qualities of, 44
curing requirements of, 43
for edge and face joints,
120-21
endurance qualities of, 44
for face-laminating sheet
goods, 132
freeze/thaw stability of, 40
as gap filler, 44
for gluing sheeet goods, 126
and health/safety issues,
41-42
for honeycomb-core
substrates, 134
initial tack of, 42
for inlay, 157
for iron-on veneering,
140-141
for joints, 150-151
for lumber-core plywood
substrates, 132
moisture content of, 37, 39
for plastic laminate, 137
removing, 162
setting speed of, 43
shelf life of, 39
solids percentage of, 39
for splined end joints, 121
for stringing, 154, 155
structural qualities of, 43
thinning, 37
tinting, 37
for torsion-box substrates,
133
Type II, 17, 35
as urea resin glue modifier,
73
for veneering, 136-37
viscosity of, 37
white vs. yellow, 34, 35
and working temperature, 42

R

Rabbit-skin glue:
conversion properties of, 10
discussed, 31

Reference boards, as gluing
accessories, 146
Repairs, glues for, 160-62, 163
Resorcinol resin glue:
application of, 68-70
assembly times for, 70
for bent laminations, 123
clamping with, 70-72
clean-up and disposal of,
66-67
composition of, 60-61
conversion properties of, 10
endurance qualities of, 72
as gap filler, 72
and health/safety issues, 62,
67-68
moisture content of, 62-63,
69
pot life of, 66
shelf life of, 63-64
solids percentage of, 63
storage temperature for, 64
structural qualities of, 72
surface sensitivity of, 62
Reversibility, of natural glues,
22
Rice-paste glue:
conversion properties of, 10
discussed, 22-23

S

Safety:
information on, 14
shop setup for, 104
Salt, as gel depressant, 29
Sheet goods:
edge and face joints with,
124
as veneer substrate, 128
Shelf life, 13
See also specific glues.
Solvents, for contact cement, 47
Spread thickness:
discussed, 109-110
gauge for, 110
Spring jointing, 118
Stringing, gluing advice for,
154-155
Substrates (for veneer):
glues for, 132-34
preparing, 129-30
sheet goods as, 128
shopmade, 128-29
solid wood as, 128
torsion boxes as, 129

T

Tagua nut, glues for, 159
Temperature, as factor in gluing
success, 101-102
Torsion boxes, as veneer
substrate, 129

U

Urea formaldehyde (UF) glue.
See Urea resin glue.
Urea resin glue:
application of, 68-70
assembly times for, 70
for bent laminations, 123
clamping with, 70-72

clean-up and disposal of,
66-67
composition of, 60-61
conversion properties of, 10
endurance qualities of, 72
as gap filler, 72
and health/safety issues,
62, 67-68
for honeycomb-core
substrates, 134
for joints, 151
for lumber-core substrates,
132
for marquetry, 157
moisture content of, 62-63,
69
one-part, mixing, 64-65
for plastic laminate, 137
pot life of, 66
PVA-modified, 73
shelf life of, 63-64
solids percentage of, 63
storage temperature for, 64
structural qualities of, 72
surface sensitivity of, 62
tints for, 62
for torsion-box substrates,
133
two-part, mixing, 64-65
for veneering, 137-38
Urea, as gel depressant, 29

V

Veneer:
as decorative material, 153
flattener for, 131
preparing, 130-31
See also Veneering.
Veneering:
glue spreader for, 135
glues for, permanent vs. non-
permanent, 136
guidelines for, 135
hammer, 138-40
iron-on, 140-41
presses for, 133
vacuum, 133-34
vacuum hammer, 140
See also Veneer.
Vinyl acrylic glue:
conversion properties of, 10
mentioned, 38
Viscosity, 12
See also specific glues.

W

Water quality, as factor in gluing
success, 103-104
Wheat-paste glue:
conversion properties of, 10
discussed, 22-23
Wood:
edge-gluing, 117-118
variability of, as factor in
gluing success, 105-106
as veneer substrate, 128

Publisher: JIM CHILDS

Associate Publisher: HELEN ALBERT

Assistant Editor: JENNIFER RENJILIAN

Publishing Coordinator: JOANNE RENNA

Technical Editor: CHRIS MINICK

Editor: RUTH DOBSEVAGE

Layout Artist: SUSAN FAZEKAS

Photographer, except where noted: WILLIAM TANDY YOUNG

Illustrator: ROBERT LaPOINTE

Typeface: PLANTIN LIGHT

Paper: G-PRINT, 68-LB.

Printer: QUEBECOR PRINTING, TENNESSEE BOOK OPERATIONS